The Observer's Pocket Series

AIRCRAFT

About the Book

The indispensable annual pocket guide to the world's latest aeroplanes and most recent versions of established aircraft types, the *Observer's Book of Aircraft* for 1981, the 30th yearly edition, embraces the newest fixed-wing aeroplanes and rotorcraft of eighteen countries. Its scope ranges from such major new civil and military débutantes as the Boeing 767 medium-range airliner, the British Aerospace 146 short-haul feederliner, Dornier's Do 228 light commuterliner, and the Siai Marchetti S.211 and Peregrine 600 trainers, all expected to commence their test programmes during this volume's year of currency, through newcomers of the past twelve months, such as the Nimrod AEW Mk 3 early warning aircraft and KC-10 Extender flight refuelling tanker, and the first representatives of a new lightweight jet trainer category, the C 22J and the Microjet, to the latest variants of a wide variety of well-established aircraft types, such as the Super 80 version of the DC-9, the interceptor variant of the Tornado and the close air support version of the MB-339.

About the Author

William Green, compiler of the *Observer's Book of Aircraft* for 30 years, is internationally known for many works of aviation reference. William Green entered aviation journalism during the early years of World War II, subsequently served with the RAF and resumed aviation writing in 1947. He is currently managing editor of one of the largest-circulation European-based aviation journals, *Air International*, and co-editor of *Air Enthusiast* and the *RAF Yearbook*.

The Observer's Book of
AIRCRAFT

COMPILED BY
WILLIAM GREEN

WITH SILHOUETTES BY
DENNIS PUNNETT

DESCRIBING 142 AIRCRAFT
WITH 247 ILLUSTRATIONS

1981 EDITION

FREDERICK WARNE

© FREDERICK WARNE & CO LTD
LONDON, ENGLAND
1981

Thirtieth edition 1981

LIBRARY OF CONGRESS CATALOG CARD NO. 57 4425

ISBN 0 7232 1618 5

Printed in Great Britain

INTRODUCTION TO THE 1981 EDITION

This edition of *The Observer's Book of Aircraft* marks its thirtieth year as an annually published reference source, and in view of the growing volume of correspondence addressed to the publishers by newer readers unfamiliar with its *raison d'être* it would seem necessary that this be restated.

The purpose of this yearly volume is to provide in compact form basic information relating to new aircraft types and variants of previously existing types that have appeared during the twelve months preceding publication, or may be expected to appear during the year of its currency, together with the latest versions of the principal aeroplanes in production today. To provide examples: the Boeing 767 is included in this edition as it is to make its début during 1981, whereas its stablemate, the Boeing 757, is excluded as flight test will not commence until next year; the MB-339 has been recorded in its current basic production training version in the four preceding annual editions, but this year appears in MB-339K single-seat dedicated close air support form in which it flew during 1980.

Apart from type or variant débutantes, the criteria adopted for inclusion of an aircraft in *The Observer's Book of Aircraft* are current production or development status; aeroplanes in process of manufacture or under active flight development at the present time. The few exceptions to these criteria are provided by those types no longer being manufactured but currently subjects of major capability-upgrade or role-change programmes (e.g., Nimrod MR Mk 2 and AEW Mk 3). The content of this volume is not, therefore, necessarily representative of the aircraft that those consulting it are most likely to see in the world's skies; such aircraft as the Boeing 707, the Concorde, the DC-8, and many others, remain today numerically among the most common types to be seen, but were, of necessity, withdrawn from the pages of the *Observer's Book* when they left the assembly lines in order to provide space for their production successors.

Nor is it possible, within the compass of this small volume, to include *all* aircraft types currently in production, although comparatively few of those that, in the compiler's view, are of major importance are excluded. However, the demands on space made by new types or new variants of earlier types dictate omission of some aeroplanes retaining production significance, and these are reinstated from time to time while their manufacture continues (e.g., the L 39 Albatros, the Bandeirants and the M-18 Dromader reappear in their latest forms in this edition).

All data is checked and, where necessary, updated in each successive year's volume, as are also the three-view silhouette drawings, these reflecting any changes introduced by the manufacturer during the course of the year, or (in the case of Soviet aircraft) the latest available information.

WILLIAM GREEN

AERITALIA G.222T

Country of Origin: Italy.

Type: General-purpose military transport.

Power Plant: Two 4,860 shp Rolls-Royce Tyne RTy 20 (Mk. 801) turboprops.

Performance: Max. cruising speed (at 54,013 lb/24 500 kg), 350 mph (563 km/h) at 20,000 ft (6 095 m); range with max. payload (15,430 lb/7 000 kg) and 10% reserves, 825 mls (1 327 km) at 20,000 ft (6 095 m), with 53 fully-equipped troops and 10% reserves, 1,060 mls (1 705 km); ferry range, 3,175 mls (5 110 km); initial climb at max. take-off weight, 2,050 ft/min (10,4 m/sec).

Weights: Operational empty, 39,682 lb (18 000 kg); max. take-off, 63,933 lb (29 000 kg).

Accommodation: Flight crew of three—four and 53 fully-equipped troops, 42 paratroops or 36 casualty stretchers plus four medical attendants.

Status: Prototype G.222T (conversion of 34th production G.222 airframe) flew for first time on May 15, 1980, with deliveries against order for 20 from Libya to commence early 1981 at one per month. Seventy-seven (all versions) laid down of which 34 built at Caselle and remainder being built at Pomigliano d'Arco, with some 45 delivered by beginning of 1981.

Notes: Variants of standard General Electric T64-P4D-powered G.222 (see 1978 edition) are the G.222VS electronic-countermeasures version and G.222RM flight inspection aircraft. The G.222T is illustrated above and opposite.

AERITALIA G.222T

Dimensions: Span, 94 ft 2 in (28,70 m); length, 74 ft 5½ in (22,70 m); height, 32 ft 1¾ in (9,80 m); wing area, 882·64 sq ft (82,00 m²).

AERMACCHI MB-339K VELTRO 2

Country of Origin: Italy.

Type: Single-seat light close air support aircraft.

Power Plant: One 4,320 lb (1 960 kg) Fiat-built Rolls-Royce Viper 632-43 turbojet.

Performance: (Combat configuration) Max. speed, 553 mph (889 km/h) at sea level; tactical radius (with four 500-lb/227-kg bombs) LO-LO-LO, 234 mls (376 km), HI-LO-HI, 403 mls (648 km); max. climb, 7,500 ft/min (38 m/sec) at sea level; time to 30,000 ft (9 145 m) from brakes release, 9·15 min.

Weights: Operational empty, 6,997 lb (3 174 kg); loaded (clean), 10,974 lb (4 978 kg); max. take-off, 13,558 lb (6 150 kg).

Armament: Two 30-mm DEFA cannon with 125 rpg. Six wing ordnance hardpoints, typical loads including four 500-lb (227-kg) bombs or six LAU-3/A launch pods each with 19 2·75-in (6,98-cm) rockets.

Status: First MB-339K (built on production jigs) flown on May 30, 1980, and components for three further examples in hand at beginning of 1981.

Notes: The MB-339K Veltro (Greyhound) 2 is a dedicated single-seat air support derivative of the MB-339 tandem two-seat basic/advanced trainer (see 1980 edition) with a new forward fuselage and increased internal fuel. The MB-339K will be built on the same production line as the two-seat MB-339, production of which will attain three monthly early in 1981 against planned procurement of 100 by the Italian Air Force and export orders (14 had been ordered by the Peruvian Air Force and 10 by the Argentine Navy by the beginning of 1981).

AERMACCHI MB-339K VELTRO 2

Dimensions: Span, 36 ft 2⅞ in (11,05 m); length, 36 ft 0 in (10,97 m); height, 12 ft 9½ in (3,90 m); wing area, 178·83 sq ft (16,61 m²).

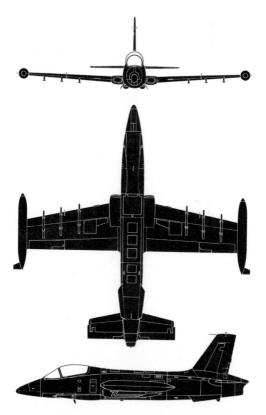

AERO L 39 ALBATROS

Country of Origin: Czechoslovakia.

Type: Tandem two-seat basic and advanced trainer (L 39Z0) weapons trainer and (L 39Z) close air support aircraft.

Power Plant: One 3,792 lb (1 720 kg) Walter Titan (licence-built Ivchenko AI-25-TL) turbofan.

Performance: (L 39C at 10,075 lb/4 570 kg) Max. speed, 485 mph (780 km/h) at 19,685 ft (6 000 m); cruise (at 9,480 lb/4 300 kg), 423 mph (680 km/h) at 16,400 ft (5 000 m); max. climb, 4,330 ft/min (22 m/sec); service ceiling, 37,730 ft (11 500 m); endurance (max. internal fuel), 2·0 hrs; max. range (with two 77 Imp gal/350 l drop tanks), 994 mls (1 600 km).

Weights: Empty equipped, 7,341 lb (3 330 kg); loaded (clean with empty wingtip tanks), 10,075 lb (4 570 kg); max. take-off, 11,618 lb (5 270 kg).

Armament: (L 39Z0) Four wing hardpoints for max. external stores load of 2,425 lb (1 100 kg), and (L 39Z) one 23-mm GSh-23 twin-barrel cannon in fuselage centreline pod.

Status: Prototype flown November 4, 1968, followed by pre-series of 10 aircraft, with series production commencing late 1972, service deliveries commencing 1974. Current production versions include L 39C basic/advanced trainer which is serving with the air forces of Czechoslovakia, Afghanistan, the DDR and the Soviet Union, the L 39Z0 weapons trainer which has been supplied to Iraq and Libya, and the L 39Z close air support version.

Notes: The L 39Z0 and L 39Z (the latter having been scheduled to enter production during 1980) feature reinforced wing and undercarriage to cater for external ordnance loads.

AERO L 39 ALBATROS

Dimensions: Span, 31 ft 0½ in (9,46 m); length, 40 ft 5 in (12,32 m); height, 15 ft 5½ in (4,72 m); wing area, 202·36 sq ft (18,80 m²).

AÉROSPATIALE TB 30 EPSILON

Country of Origin: France.
Type: Tandem two-seat primary-basic trainer.
Power Plant: One 300 hp Avco Lycoming IO-540-L1B5-D six-cylinder horizontally opposed engine.
Performance: (Estimated) Max. speed, 218 mph (352 km/h) at sea level; max. cruise (75% power), 207 mph (333 km/h) at 6,000 ft (1 830 m); max. climb, 1,800 ft/min (9,14 m/sec); service ceiling, 20,000 ft (6 100 m); endurance, 3·8 hrs.
Weights: Empty equipped, 1,825 lb (828 kg); max. take-off, 2,590 lb (1 175 kg).
Status: First prototype flown on December 22, 1979, with second following on July 12, 1980. Both prototypes subsequently modified and flight testing resumed with first of these on October 31, 1980. An order for an initial production batch of 30 Epsilons for the *Armée de l'Air* was anticipated early 1981 against eventual requirement for approximately 150 aircraft.
Notes: The Epsilon programme was lauched by the *Armée de l'Air* in March 1978, but unacceptable pitch/yaw coupling characteristics demonstrated by the prototypes resulted, during the course of 1980, in the application of extended, rounded and upswept wingtips, redesign of the rear fuselage and vertical tail surfaces, the lowering of the horizontal tail and the addition of a ventral fin. The Epsilon is to be employed by the *Armée de l'Air* for the first 80—100 hours of flying training, the cockpit design, manoeuvre performance and landing speed being matched to the Alpha Jet to which pupils will progress.

AÉROSPATIALE TB 30 EPSILON

Dimensions: Span, 25 ft 11½ in (7,92 m); length, 24 ft 10½ in (7,59 m).

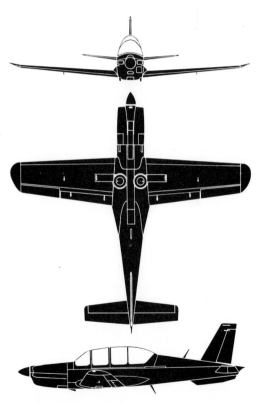

AIRBUS A300B4-100

Country of Origin: International consortium.
Type: Medium-haul commercial transport.
Power Plant: (A300B4-101) Two 51,000 lb (23 130 kg)
General Electric CF6-50C turbofans.
Performance: Max. cruise, 578 mph (930 km/h) at 28,000 ft
(9 185 m); econ. cruise, 540 mph (869 km/h) at 31,000 ft
(9 450 m); long-range cruise, 521 mph (839 km/h) at 33,000 ft
(10 060 m); range (with max. payload, no reserve), 2,618 mls
(4 213 km); max. range (with 48,350-lb/21 932-kg payload),
3,994 mls (6 428 km).
Weights: Operational empty, 194,130 lb (85 060 kg); max.
take-off, 347,200 lb (157 500 kg).
Accommodation: Crew of three on flight deck with provision
for two-man operation. Seating for 220–336 passengers in
main cabin in six-, seven- or eight-abreast layouts.
Status: First and second A300Bs flown October 28, 1972,
and February 5, 1973 (B1 standard), respectively, with third
(to B2 standard) flying on June 28, 1973. First A300B4 flown
December 26, 1974, and at the beginning of 1981, production rate was 3·0 per month against firm orders for 227
(plus 89 options), and scheduled to rise to eight aircraft
monthly in 1984.
Notes: The A300B is manufactured by a consortium of Aérospatiale, British Aerospace and Deutsche Airbus. The A300B2-
100 is the basic version, the A300B2-200 having Krueger flaps
for improved field performance, and the A300B4-100 is a
longer-range model with the Krueger flaps and a centre-section
fuel tank, the A300B4-200 having increased gross weight
(363,800 lb/165 000 kg). The A310 is a short-fuselage, re-
winged derivative scheduled to fly in 1982 with service entry
in 1983 and for which 76 orders (plus 70 options) had been
placed by the beginning of 1981. The A300B4-600 (with new-
generation engines and some A310 features) is expected to
fly in July 1983 (first deliveries to Saudi Arabian Airlines).

AIRBUS A300B4-100

Dimensions: Span, 147 ft 1¼ in (44,84 m) : length, 175 ft 11 in (53,62 m) ; height, 54 ft 2 in (16,53 m) ; wing area, 2,799 sq ft (260,00 m²)

ANTONOV AN-28 (CASH)

Country of Origin: USSR.

Type: Light STOL general-purpose transport and feederliner.

Power Plant: Two 960 shp Glushenkov TVD-10B turbo-props.

Performance: Max. continuous cruising speed, 217 mph (350 km/h); range (with 3,415-lb/1 550-kg payload), 620 mls (1 000 km), (with max. fuel), 805 mls (1 300 km); initial climb rate, 2,360 ft/min (11,99 m/sec).

Weights: Normal loaded, 12,785 lb (5 800 kg); max. take-off, 13,450 lb (6 100 kg).

Accommodation: Flight crew of one or two and up to 18 passengers in high-density configuration, but standard configuration for 15 passengers in three-abreast seating (one to port and two to starboard), the seats folding back against the cabin walls when the aircraft is employed in the freighter or mixed passenger/freight roles. Alternative versions provide for six or seven passengers in an executive transport arrangement and an aeromedical version accommodates six stretchers and a medical attendant.

Status: Initial prototype flown as An-14M in September 1969. A production prototype was tested early in 1974 with 810 hp Isotov TVD-850 turboprops, the aircraft having meanwhile been redesignated An-28, and this was re-engined with Glushenkov TVD-10A engines with which it first flew in April 1975. No series production has been undertaken in the Soviet Union, but licence production is being initiated in Poland by PZL which is expected to commence production deliveries in 1983–84, current planning calling for export of 1,200 to the Soviet Union by 1990.

16

ANTONOV AN-28 (CASH)

Dimensions: Span, 72 ft 2⅛ in (22,06 m); length, 42 ft 6⅞ in (12,98 m); height, 15 ft 1 in (4,60 m); wing area, 433·58 sq ft (40,28 m²).

ANTONOV AN-72 (COALER)

Country of Origin: USSR.
Type: Short-haul STOL transport.
Power Plant: Two 14,330 lb (6 500 kg) Lotarev D-36 turbofans.
Performance: Max. cruising speed, 447 mph (720 km/h); range with max. payload (16,534 lb/7 500 kg) and 30 min reserves, 620 mls (1 000 km), with max. fuel, 1,990 mls; normal operating altitude, 26,250–32,800 ft (8 000–10 000 m).
Weights: Loaded (for 3,280-ft/1 000-m runway), 58,420 lb (26 500 kg); max. take-off, 67,240 lb (30 500 kg).
Accommodation: Flight crew of two–three and up to 32 passengers on fold-down seats along cabin sides, or 24 casualty stretchers plus one medical attendant. Rear loading ventral ramp with clamshell doors for cargo hold which has overhead hoist and can be provided with roller floor.
Status: First of two prototypes flown on December 22, 1977. No production plans had been revealed at closing for press.
Notes: Similar in concept to the Boeing YC-14 (see 1978 edition) in utilising upper-surface-blowing, engine exhaust gases flowing over the upper wing surfaces and the inboard double-slotted flaps, the An-72 has been designed primarily for use in remote areas and is capable of operating for short, semi-prepared airstrips. Commercial use is likely to be restricted by high operating costs to areas inaccessible to more conventional aircraft and the intended role of the An-72 would appear to be primarily military.

18

ANTONOV AN-72 (COALER)

Dimensions: Span, 84 ft 9 in (25,83 m); length, 87 ft 2½ in (26,58 m); height, 27 ft 0 in (8,24 m).

BAe ONE ELEVEN 500

Country of Origin: United Kingdom.
Type: Short- to medium-haul commercial transport.
Power Plant: Two 12,550 lb (5 698 kg) Rolls-Royce Spey 512 DW turbofans.
Performance: Max. cruise, 541 mph (871 km/h) at 21,000 ft (6 400 m); econ cruise, 461 mph (742 km/h) at 25,000 ft (7 620 m); range with capacity payload (27,090 lb/12 286 kg) and reserves, 1,705 mls (2 744 km), with max. fuel and reserves, 2,165 mls (3 484 km).
Weights: Operational empty, 53,911 lb (24 454 kg); max. take-off, 99,650–104,500 lb (45 200–47 400 kg).
Accommodation: Flight crew of two and up to 119 passengers in main cabin.
Status: Prototype Series 500 (converted from Series 400 development aircraft) flown June 30, 1967, with first production aircraft following on February 7, 1968. Total of 245 ordered (all versions) with Series 475 and Series 500 production in process of transfer to Rumania. One of former and two of latter for Rumania to be followed by 21 complete sets of components (both series) for assembly by CNIAR in Rumania by 1985. Thereafter, CNIAR is to produce further 60 Series 475s and 500s at rate of six annually.
Notes: The One-Eleven first flew on August 20, 1963, production models including the physically similar Series 200 and 300 with Spey 506s and Spey 511s respectively, the Series 400 for US operation, the Series 475 combining the fuselage and accommodation of the Series 400 with similar redesigned wing and uprated engines to those of the Series 500 which introduced a lengthened fuselage and wingtip extensions.

BAe ONE-ELEVEN 500

Dimensions: Span, 93 ft 6 in (28,50 m); length, 107 ft 0 in (32,61 m); height, 24 ft 6 in (7,47 m); wing area, 10,031 sq ft (95,78 m²).

21

BAe 146

Country of Origin: United Kingdom.
Type: Short-haul feederliner.
Power Plant: Four 6,700 lb (3 040 kg) Avco Lycoming ALF 502R-3 turbofans.
Performance: (146-100 estimated) Max. cruising speed, 490 mph (788 km/h) at 22,000 ft (6 705 m); econ. cruise, 440 mph (709 km/h) at 30,000 ft (9 145 m); long-range cruise, 427 mph (687 km/h) at 30,000 ft (9 145 m); range (max. payload), 748 mls (1 204 km), (with 71 passengers), 1,128 mls (1 816 km).
Weights: Operational empty, 44,570 lb (20 217 kg); max. take-off, 73,850 lb (33 500 kg).
Accommodation: Flight crew of two and maximum seating for 88 passengers six-abreast or 76 five-abreast.
Status: First of three 146-100s assigned to flight test programme scheduled to fly May 1981, and first 146-200 (eighth aeroplane off line) to fly March 1982. First delivery (to LAPA) September 1982.
Notes: The BAe 146, for which the first orders (from airlines in Argentina and USA) were placed in 1980, is currently being built in two versions, the -100, described and illustrated above, and the stretched -200 illustrated opposite. The latter has an overall length of 93 ft 8½ in (28,56 m) and will accommodate up to 110 passengers. The BAe 146 is optimised to operate over stage lengths of about 150 miles (240 km) with unrefuelled multi-stop capability. The BAe 146-100 has an overall length of 85 ft 10 in (26,16 m) but is similar to the -200 in all other respects.

BAE 146-200

Dimensions: Span, 85 ft 5 in (26,42 m); length, 93 ft 8½ in (28,56 m); height, 28 ft 3 in (8,61 m); wing area, 832 sq ft (77,30 m²).

BAe HS 125-700

Country of Origin: United Kingdom.

Type: Light business executive transport.

Power Plant: Two 3,700 lb (1 680 kg) Garrett AiResearch TFE 731-3-1H turbofans.

Performance: High-speed cruise, 495 mph (796 km/h); long-range cruise, 449 mph (722 km/h); range (with 1,200-lb/544-kg payload and 45 min reserve), 2,705 mls (4 355 km); time to 35,000 ft (10 675 m), 19 min; operating altitude, 41,000 ft (12 500 m).

Weights: Typical basic, 13,327 lb (6 045 kg); max. take-off, 24,200 lb (10 977 kg).

Accommodation: Normal flight crew of two and basic layout for eight passengers, with alternative layouts for up to 14 passengers.

Status: Series 700 development aircraft flown June 28, 1976, followed by first production aircraft on November 8, 1976. Some 110 HS 125-700s had been delivered by the beginning of 1981, when production rate was three per month. The 500th HS 125 sale (of which 142 were -700s) was announced in September 1980.

Notes: The HS 125-700 differs from the -600 (see 1976 edition) that it supplanted primarily in having turbofans in place of Viper 601 turbojets and various aerodynamic improvements. A programme has been established for the conversion of earlier HS 125s for the Garrett AiResearch turbofans, some 50 aircraft having been committed to this retrofit programme by the beginning of 1981. Various options were under study at the beginning of 1981 for developed versions, ranging from a modest fuselage stretch and 4,000 lb (1 814 kg) TFE 731-5 engines to an entirely new wing and new engines.

24

BAE HS 125-700

Dimensions: Span, 47 ft 0 in (14,32 m); length, 50 ft 8½ in (15,46 m); height, 17 ft 7 in (5,37 m); wing area, 353 sq ft (32,80 m²).

BAe HS 748-2B

Country of Origin: United Kingdom.

Type: Short- to medium-range commercial and military transport.

Power Plant: Two 2,280 ehp Rolls-Royce Dart RDa 7 Mk 536-2 turboprops.

Performance: Max. cruise speed, 283 mph (456 km/h) at 12,000 ft (3 660 m); long-range cruise, 269 mph (433 km/h) at 25,000 ft (7 620 m); range (with max. payload—12,500 lb/5 670 kg), 1,324 mls (2 130 km) at 279 mph (448 km/h) at 17,000 ft (5 180 m), (max. fuel and 8,980-lb/4 073-kg payload), 2,050 mls (3 300 km) at 269 mph (433 km/h) at 25,000 ft (7 620 m).

Weights: Operational empty, 26,000 lb (11 794 kg); max. take-off, 46,500 lb (21 092 kg).

Accommodation: Normal flight crew of two with arrangements for 44 to 58 passengers four abreast in paired seats with central aisle.

Status: First prototype 748 flown June 24, 1960, and prototype Series 2B flown on June 22, 1979, this version superseding the Series 2A with first delivery (to Air Madagascar) in January 1980. Nine HS 748-2Bs delivered during 1980, with anticipated production of 12 in 1981 and 18 in 1982. Orders for HS 748 series aircraft of all types (including Andovers) total 350 at the beginning of 1981.

Notes: The HS 748-2B differs from the -2A which it succeeds in having extended wingtips, marginally uprated engines, noise and drag reduction modifications, new fuel management system and modernised cockpit. For the military freighter rôle a structurally strengthened floor and enlarged freight door (as shown opposite) are options.

26

BAe HS 748-2B

Dimensions: Span, 102 ft 6 in (31,23 m); length, 67 ft 0 in (20,42 m); height, 24 ft 10 in (7,57 m); wing area, 828·87 sq ft (77,00 m²).

BAe JETSTREAM 31

Country of Origin: United Kingdom.
Type: Light business and utility transport.
Power Plant: Two 940 shp Garrett AiResearch TPE 331-10 turboprops.
Performance: Max. cruising speed, 303 mph (488 km/h) at 16,000 ft (4 875 m); range cruise, 291 mph (469 km/h); initial climb, 2,231 ft/min (11,33 m/sec); service ceiling, 31,600 ft (9 480 m); range (with 30 min reserves plus 5%—six passengers), 1,275 mls (2 053 km), (eight passengers), 1,150 mls (1 852 km), (18 passengers), 483 mls (778 km).
Weights: Empty, 7,606 lb (3 450 kg); max. take-off, 14,110 lb (6 400 kg).
Accommodation: Two seats side-by-side on flight deck with 8–10 passengers in main cabin of corporate executive version and optional commuter arrangements for 18 passengers in two-plus-one seating.
Status: Flight development Jetstream 31 (converted from a Series 1 airframe) flown March 28, 1980, with first production aircraft scheduled for second half of 1981. Production rate of 25 aircraft annually proposed for 1982, but production launch decision awaited at beginning of 1981.
Notes: The Jetstream 31 is derived from the Handley Page H.P. 137 Jetstream by the Scottish Division of British Aerospace, the original Jetstream prototype having flown on August 18, 1967. The Jetstream 31 is currently being offered in corporate, commuter and military versions, and consideration has been given to a "stretched" version accommodating up to 25 passengers plus a cabin attendant.

28

BAe JETSTREAM 31

Dimensions: Span, 52 ft 0 in (15,85 m); length, 47 ft 1½ in (14,36 m); height, 10 ft 6 in (3,20 m); wing area, 270 sq ft (25,08 m²).

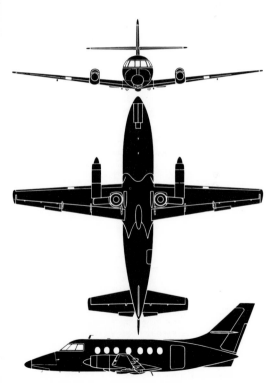

BAe HARRIER G.R. Mk 3

Country of Origin: United Kingdom.

Type: Single-seat V/STOL strike and reconnaissance fighter.

Power Plant: One 21,500 lb (9 760 kg) Rolls-Royce Pegasus 103 vectored-thrust turbofan.

Performance: Max. speed, 720 mph (1 160 km/h) or Mach 0·95 at 1,000 ft (305 m), with typical external ordnance load, 640–660 mph (1 030–1 060 km) or Mach 0·85–0·87 at 1,000 ft (305 m); cruise, 560 mph (900 km/h) or Mach 0·8 at 20,000 ft (6 096 m); tactical radius for HI-LO-HI mission, 260 mls (418 km), with two 100 Imp gal (455 l) external tanks, 400 mls (644 km).

Weights: Empty, 12,400 lb (5 624 kg); max. take-off (VTO), 18,000 lb (8 165 kg); max. take-off (STO), 23,000+ lb (10 433+ kg); approx. max. take-off, 26,000 lb (11 793 kg).

Armament: Provision for two 30-mm Aden cannon with 130 rpg and up to 5,000 lb (2 268 kg) of ordnance.

Status: First of six pre-production aircraft flown August 31, 1966, with first of 77 G.R. Mk 1s for RAF following December 28, 1967. Production of G.R. Mk 1s and 13 T. Mk 2s (see 1969 edition) for RAF completed. Production of 102 Mk 50s (equivalent to G.R. Mk 3) and eight Mk 54 two-seaters (equivalent to T. Mk 4) for US Marine Corps, and six Mk 50s and two Mk 54s ordered (via the USA) by Spain (by which known as Matador), plus follow-on orders for 13 G.R. Mk 3s and four T. Mk 4s also completed. Production continuing in 1981 against follow-on order for 24 Mk 3s for the RAF, plus one Mk 4 for the Royal Navy and two for the Indian Navy, and five Mk 50s for Spain.

Notes: RAF Harriers have been progressively brought up to G.R. Mk 3 and T. Mk 4 standards by installation of Pegasus 103 similar to that installed in Mk 50 (AV-8A) for USMC.

BAe HARRIER G.R. Mk 3

Dimensions: Span, 25 ft 3 in (7,70 m); length, 45 ft 7¾ in (13,91 m); height, 11 ft 3 in (3,43 m); wing area, 201·1 sq ft (18,68 m²).

BAe HAWK T. Mk 1

Country of Origin: United Kingdom.

Type: Two-seat multi-purpose trainer and light tactical aircraft.

Power Plant: One 5,340 lb (2 422 kg) Rolls-Royce Turbo-méca RT.172-06-11 Adour 151 turbofan.

Performance: Max. speed, 622 mph (1 000 km/h) at sea level or Mach 0·815, 580 mph (933 km/h) at 36,000 ft (10 970 m) or Mach 0·88; radius of action (HI-LO-HI profile with 5,600-lb/2 540-kg weapons load), 345 mls (560 km), (with 3,000-lb/1 360-kg weapons load and two 100 Imp gal/455 l drop tanks), 645 mls (1 040 km); time to 30,000 ft (9 145 m), 6·1 min; service ceiling, 48 000 ft (14 630 m).

Weights: Empty, 8,040 lb (3 647 kg); loaded (clean), 11,100 lb (5 040 kg); max. take-off, 17,085 lb (7 757 kg).

Armament: (Weapon training) One fuselage centreline and two wing stores stations each stressed for 1,120 lb (508 kg), and (attack) two additional similarly-stressed wing stations. Max. external stores load of 5,600 lb (2 540 kg).

Status: Single pre-production example flown August 21, 1974, first production example flown May 19, 1975, and some 150 delivered by beginning of 1981 against RAF orders for 175 (with repeat order for 18 pending). Fifty ordered by Finland (Mk 51), 12 by Kenya (Mk 52) and eight by Indonesia (Mk 53). Forty-six of those ordered by Finland are being assembled by Valmet from component sets. Deliveries against all three export orders were in process at the beginning of 1981.

Notes: At the beginning of 1981, it was anticipated that kits would be procured to convert approximately half of the RAF's Hawks to carry a pair of AIM-9 Sidewinder AAMs on the inboard wing pylons so that the aircraft can supplement the United Kingdom's defence force in an emergency, flying from operational bases and piloted by weapons instructors.

BAe HAWK T. Mk 1

Dimensions: Span, 30 ft 9¾ in (9,39 m); length, 38 ft 10⅔ in (11,85 m); height, 13 ft 1 in (4,00 m); wing area, 179·64 sq ft (16,69 m²).

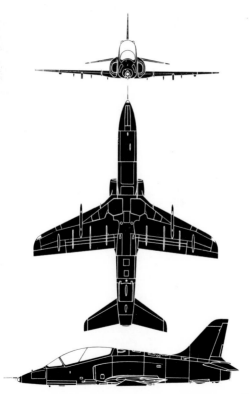

BAe NIMROD M.R. Mk 2

Country of Origin: United Kingdom.
Type: Long-range maritime patrol aircraft.
Power Plant: Four 12,160 lb (5 515 kg) Rolls-Royce RB. 168-20 Spey Mk 250 turbofans.
Performance: Max. speed, 575 mph (926 km/h); max. transit speed, 547 mph (880 km/h); econ. transit speed, 490 mph (787 km/h); typical ferry range, 5,180–5,755 mls (8 340–9 265 km); typical endurance, 12 hrs.
Weights: Max. take-off, 177,500 lb (80 510 kg); max. overload, 192,000 lb (87 090 kg).
Armament: Ventral weapons bay accommodating full range of ASW weapons (Stingray homing torpedoes, mines, depth charges, etc). Provision for two underwing pylons on each side for total of four Aérospatiale AS. 12 ASMs.
Accommodation: Normal operating crew of 12 with two pilots and flight engineer on flight deck and nine navigators and sensor operators in tactical compartment.
Status: First of 38 Nimrod M.R. Mk 1s flown on June 28, 1968. Completion of this batch in August 1972 followed by delivery of three Nimrod R. Mk 1s for special electronics reconnaissance, and eight more M.R. Mk 1s ordered in 1973. Thirty-two Nimrod M.R. Mk 1s are being progressively brought up to M.R. Mk 2 standard in a programme to continue to mid-1984, and first accepted by RAF on August 23, 1979.
Notes: Nimrod M.R. Mk 2 possesses 60 times more computer power than the M.R. Mk 1, is equipped with the advanced Searchwater maritime radar, an AQS-901 acoustics system compatible with the Barra sonobuoy, and will have EWSM (Electronic Warfare Support Measures) wingtip pods.

BAe NIMROD M.R. Mk 2

Dimensions: Span, 114 ft 10 in (35,00 m); length, 126 ft 9 in (38,63 m); height, 29 ft 8½ in (9,01 m); wing area, 2,121 sq ft (197,05 m²).

BAe NIMROD A.E.W. Mᴋ 3

Country of Origin: United Kingdom.

Type: Airborne warning and control system aircraft.

Power Plant: Four 12,160 lb (5 515 kg) Rolls-Royce RB. 168-20 Spey Mk 250 turbofans.

Performance: No details have been released for publication, but maximum and transit speeds are likely to be generally similar to those of the M.R. Mk 2, and maximum endurance is in excess of 10 hours. Mission requirement calls for 6–7 hours on station at 29,000–35,000 ft (8 840–10 670 m) at approx. 350 mph (563 km/h) at 750–1,000 mls (1 120–1 600 km) from base.

Weights: No details available.

Accommodation: Flight crew of four and tactical team of six. Tactical team comprises tactical air control officer, communications control officer, EWSM (Electronic Warfare Support Measures) operator and three air direction officers located in the tactical area of the cabin.

Status: Total of 11 Nimrod M.R. Mk 1 airframes being rebuilt to A.E.W. Mk 3 standard of which fully representative prototype flew on July 16, 1980. Two additional A.E.W. Mk 3 aircraft involved in development programme early 1981, with RAF service phase-in commencing in 1982.

Notes: The Nimrod A.E.W. Mk 3 airborne warning and control system aircraft is equipped with Marconi mission system avionics with identical radar aerial mounted in nose and tail, these being synchronised and each sequentially sweeping through 180 deg in azimuth and providing uninterrupted coverage throughout 360 deg of combined sweep. EWSM pods are located at the wingtips and weather radar is installed in the front of the starboard wing pinion tank.

BAE NIMROD A.E.W. Mk 3

Dimensions: Span, 115 ft 1 in (35,08 m); length, 137 ft 8½ in (41,97 m); height, 35 ft 0 in (10,67 m); wing area, 2,121 sq ft (197,05 m²).

BAe SEA HARRIER F.R.S. MK 1

Country of Origin: United Kingdom.
Type: Single-seat V/STOL shipboard multi-role fighter.
Power-Plant: One 21,500 lb (9 760 kg) Rolls-Royce Pegasus 104 vectored-thrust turbofan.
Performance: (Estimated) Max. speed, 720 mph (1 160 km/h) at 1,000 ft (305 m) or Mach 0·95, with two Martel ASMs and two Sidewinder AAMs, 640–660 mph (1 030–1 060 km/h) or Mach 0·85–0·87; tactical radius (intercept mission with two 100 Imp gal/455 l drop tanks, two 30-mm cannon and two Sidewinder AAMs), 450 mls (725 km), (strike mission HI-LO-HI profile), 330 mls (480 km).
Weights: Empty, 12,500 lb (5 670 kg); max. STO take-off, 22,500 lb (10 206 kg); max. overload, 25,000 lb (11 339 kg).
Armament: Provision for two (flush-fitting) podded 30-mm Aden cannon with 130 rpg beneath fuselage. Five external hardpoints (one fuselage and four wing) each stressed for 1,000 lb (453,5 kg), with max. external ordnance load for STO (excluding cannon) of 5,000 lb (2 268 kg). Typical loads include two Martel or Harpoon ASMs on inboard wing pylons and two Sidewinder AAMs on outboard pylons.
Status: First Sea Harrier (built on production tooling) flown on August 21, 1978, with deliveries against 34 ordered for Royal Navy commencing in second half of 1979. Six Sea Harriers were ordered late 1979 for delivery to Indian Navy.
Notes: First Sea Harrier squadron, No 700A, was formally commissioned on September 19, 1979.

BAe SEA HARRIER F.R.S. Mk 1

Dimensions: Span, 25 ft 3 in (7,70 m); 47 ft 7 in (14,50 m); height, 12 ft 2 in (3,70 m); wing area, 201·1 sq ft (18,68 m²).

BEECHCRAFT SKIPPER 77

Country of Origin: USA.

Type: Side-by-side two-seat primary trainer.

Power Plant: One 115 hp Avco Lycoming O-235-L2C four-cylinder horizontally-opposed engine.

Performance: Cruise at 4,500 ft (1 372 m), 121 mph (195 km/h) at 80% power, 112 mph (180 km/h) at 65% power, 107 mph (172 km/h) at 59% power, at 8,500 ft (2 590 m), 110 mph (177 km/h) at 61% power, 105 mph (169 km/h) at 55% power; initial climb, 720 ft/min (3,7 m/sec); service ceiling, 12,900 ft (3 932 m); range (with reserves), 376 mls (605 km) at 80% power at 4,500 ft (1 372 m), 447 mls (719 km) at 8,500 ft (2 590 m) at 61% power.

Weights: Empty, 1,103 lb (500 kg); max. take-off, 1,675 lb (760 kg).

Status: Prototype flown on February 6, 1975, with production prototype following September 1978. Production deliveries commenced April 1979, and approximately 12 per month were being built at the beginning of 1981 when some 200 had been delivered.

Notes: Evolved as a low-cost primary trainer placing emphasis on simplicity of maintenance and low operating cost, the Skipper 77 utilises a NASA-developed high-lift GA(W)-1 wing of tubular-spar concept, and the flaps and ailerons are actuated by torque tubes rather than a conventional cable-and-pulley system. The Skipper closely resembles the competitive Piper PA-38 Tomahawk (see pages 180–181).

BEECHCRAFT SKIPPER 77

Dimensions: Span, 30 ft 0 in (9,14 m); length, 24 ft 0 in (7,32 m); height, 6 ft 11 in (2,11 m); wing area, 129·8 sq ft (12,06 m²).

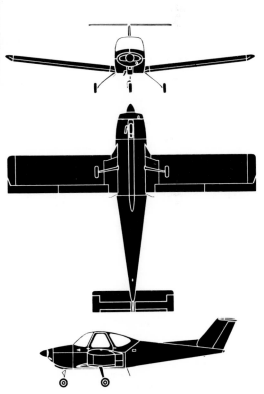

BEECHCRAFT DUCHESS 76

Country of Origin: USA.
Type: Light cabin monoplane.
Power Plant: Two 180 hp Avco Lycoming O-360-A1G6D six-cylinder horizontally-opposed engines.
Performance: Max. speed, 197 mph (317 km/h); max. cruise (at 3,600 lb/634 kg), 191 mph (307 km/h) at 6,000 ft (1 830 m); normal cruise, 176 mph (283 km/h) at 10,000 ft (3 050 m); econ. cruise, 172 mph (277 km/h) at 12,000 ft (3 658 m); range at econ. cruise (45 min reserves), 898 mls (1 445 km); initial climb, 1,248 ft/min (6,3 m/sec).
Weights: Empty, 2,460 lb (1 116 kg); max. take-off, 3,900 lb (1 770 kg).
Accommodation: Pilot and three passengers in individual seats, with provision for up to 180 lb (81,6 kg) of baggage in separate compartment.
Status: Prototype flown September 1974, production being initiated in the spring of 1977, and the first production example flying on May 24, 1977. First deliveries were made early 1978, and some 400 had been delivered by the beginning of 1981.
Notes: Bearing a close resemblance to the Piper PA-44 Turbo Seminole (see 1980 Edition), the Duchess embodies handed propellers and honeycomb-bonded wings.

BEECHCRAFT DUCHESS 76

Dimensions: Span, 38 ft 0 in (11,58 m); length, 29 ft 0 in (8,84 m); height, 9 ft 6 in (2,89 m); wing area, 181 sq ft (16,81 m²).

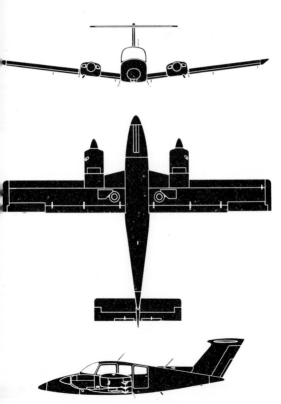

BEECHCRAFT T-34C (TURBINE MENTOR)

Country of Origin: USA.

Type: Tandem two-seat primary trainer.

Power Plant: One 680 shp (derated to 400 shp) Pratt & Whitney (Canada) PT6A-25 turboprop.

Performance: Max. cruise, 213 mph (343 km/h) at sea level, 239 mph (384 km/h) at 10,000 ft (3 050 m); range (5% and 20 min reserve), 787 mls (1 265 km) at 220 mph (354 km/h) at 17,500 ft (5 340 m), 915 mls (1 470 km) at 222 mph (357 km/h) at 25,000 ft (7 625 m); initial climb, 1,430 ft/min (7,27 m/sec).

Weights: Empty equipped, 3,015 lb (1 368 kg); normal loaded, 4,249 lb (1 927 kg).

Status: First of two YT-34Cs flown September 21, 1974, and production continuing at beginning of 1981 for US Navy which has total requirement for some 278. Export T-34C-1 was delivered to Algeria, Argentina (Navy), Ecuador, Indonesia, Morocco and Peru during 1978.

Notes: Updated derivative of Continental O-470-13-powered Model 45, the T-34C is fitted with a torque-limited PT6A-25 turboprop affording 400 shp, but the T-34C-1 may be fitted with a version of the PT6A-25 derated to 550 shp, wing racks for external ordnance an an armament control system to permit operation as an armament trainer or light-counter-insurgency aircraft. With a max. take-off weight of 5,425 lb (2 460 kg), the T-34C-1 has two 600-lb (272-kg) capacity wing inboard stores stations and two 300-lb (136-kg) capacity outboard stations.

BEECHCRAFT T-34C (TURBINE MENTOR)

Dimensions: Span, 33 ft 4¾ in (10,18 m); length, 28 ft 8½ in (8,75 m); height, 9 ft 10⅞ in (3,02 m); wing area, 179·56 sq ft (16,68 m²).

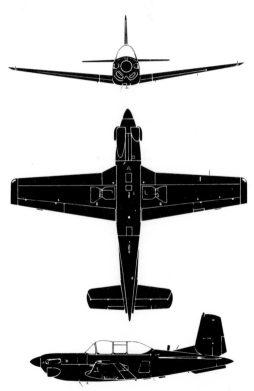

BOEING MODEL 727-200

Country of Origin: USA.

Type: Short- to medium-range commercial transport.

Power Plant: Three 14,500 lb (6 577 kg) Pratt & Whitney JT8D-9 turbofans (with 15,000 lb/6 804 kg JT8D-11s or 15,500 lb/7 030 kg JT8D-15s as options).

Performance: Max. speed, 621 mph (999 km/h) at 20,500 ft (6 250 m); max. cruise, 599 mph (964 km/h) at 24,700 ft (7 530 m); econ. cruise, 570 mph (917 km/h) at 30,000 ft (9 145 m); range with 26,400-lb (11 974-kg) payload and normal reserves, 2,850 mls (4 585 km), with max. payload (41,000 lb/18 597 kg), 1,845 mls (2 970 km).

Weights: Operational empty (basic), 97,525 lb (44 235 kg), (typical), 99,000 lb (44 905 kg); max. take-off, 208,000 lb (94 347 kg).

Accommodation: Crew of three on flight deck and six-abreast seating for 163 passengers in basic arrangement with max. seating for 189 passengers.

Status: First Model 727-100 flown February 9, 1963, with first delivery (to United) following October 29, 1963. Model 727-200 flown July 27, 1967, with first delivery (to Northeast) on December 11, 1967. Deliveries from mid-1972 have been of the so-called "Advanced 727-200" (to which specification refers) and sales of Model 727s have reached 1,811 at the beginning of 1981, with 1,695 delivered and production running at 8·75 aircraft monthly with 105 to be delivered during 1981.

Notes: The Model 727-200 is a "stretched" version of the 727-100 (see 1972 edition). Deliveries of the "Advanced 727" with JT8D-17 engines of 16,000 lb (7 257 kg), permitting an increase of 3,500 lb (1 587 kg) in payload, began (to Mexicana) in June 1974.

BOEING MODEL 727-200

Dimensions: Span, 108 ft 0 in (32,92 m); length, 153 ft 2 in 46,69 m); height, 34 ft 0 in (10,36 m); wing area, 1,560 sq ft (144,92 m²).

BOEING MODEL 737-200

Country of Origin: USA.

Type: Short-haul commercial transport.

Power Plant: Two 16,000 lb (7 258 kg) Pratt & Whitney JT8D-17 turbofans.

Performance: Max. cruising speed, 564 mph (908 km/h) at 25,000 ft (7 620 m); long-range cruise, 481 mph (775 km/h) at 35,000 ft (10 670 m); econ. cruise, 502 mph (808 km/h) at 33,000 ft (10 055 m); max. range (with 21,750-lb/9 866-kg payload), 3,086 mls (4 967 km), (with max. payload of 34,000 lb/15 422 kg), 1,750 mls (2 817 km).

Weights: Operational empty, 61,210 lb (27 764 kg); max. take-off, 117,000 lb (53 071 kg).

Accommodation: Flight crew of two and up to 130 passengers in six-abreast seating with alternative arrangement for 115 passengers.

Status: Model 737 initially flown on April 9, 1967, with first deliveries (737-100 to Lufthansa) same year. Stretched 737-200 flown on August 8, 1967, with deliveries (to United) in following year. Total orders for 867 (including 19 -200s to USAF as T-43A navigational trainers—see 1975 edition) by beginning of 1981, with 710 delivered. Production rising from 94 in 1980 to 107 in 1981.

Notes: At the beginning of 1981, it was anticipated that work would proceed on a further "stretched" version of the basic aircraft, the Model 737-300 for delivery early 1985. The fuselage will be stretched 84 in (213 cm), and the wing will be strengthened and slightly extended, and power will be provided by new-generation engines. The 100th customer for the Model 737 was announced in September 1980, and production of this type is expected to continue throughout the present decade.

BOEING MODEL 737-200

Dimensions: Span, 93 ft 0 in (28,35 m); length, 100 ft 0 in (30,48 m); height, 37 ft 0 in (11,28 m); wing area, 980 sq ft (91,05 m²).

BOEING MODEL 747-200B

Country of Origin: USA.
Type: Long-range large-capacity commercial transport.
Power Plant: Four 47,000 lb (21 320 kg) Pratt & Whitney JT9D-7W turbofans.
Performance: Max. speed at 600,000 lb (272 155 kg), 608 mph (978 km/h) at 30,000 ft (9 150 m); long-range cruise, 589 mph (948 km/h) at 35,000 ft (10 670 m); range with max. fuel and FAR reserves, 7,080 mls (11 395 km), with 79,618-lb 36 114-kg) payload, 6,620 mls (10 650 km); cruising ceiling 45,000 ft (13 715 m).
Weights: Operational empty, 361,216 lb (163 844 kg); max. take-off, 775,000 lb (351 540 kg).
Accommodation: Normal flight crew of three and basic accommodation for 66 first-class and 308 economy-class passengers. Alternative layouts for 447 or 490 economy-class passengers nine- and 10-abreast respectively.
Status: First Model 747-100 flown on February 9, 1969, the first Model 747-200 (88th aircraft off the line) following on October 11, 1970, and the first Model 747SP flying on July 4, 1975. Orders (all versions and including six E-4 command posts for USAF—see 1980 edition) totalled 564 at beginning of 1981, with 488 delivered 2nd 63 for delivery in 1981.
Notes: Principal versions of the Model 747 are the -100 and -200, the latter with increased fuel and take-off weights, these being available (in current -100B and -200B forms) with a wide variety of engines. In addition, there are the lighter-weight shorter-bodied 747SP (see 1980 edition) and shorter-range 747SR derivatives of the 747-100B, and the 747-200C convertible passenger/freighter and 747-200F freighter derivatives of the 747-200B.

BOEING MODEL 747-200B

Dimensions: Span, 195 ft 8 in (59,64 m); length, 231 ft 4 in (70,51 m); height, 63 ft 5 in (19,33 m); wing area, 5,685 sq ft (528,15 m²).

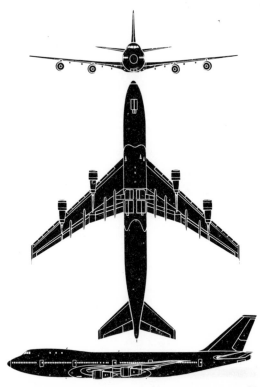

BOEING 767-200

Country of Origin: USA.

Type: Medium-range commercial transport.

Power Plant: Two 47,700 lb (21 656 kg) Pratt & Whitney JT9D-7R4D turbofans, or 44,300 lb (20 112 kg) JT9D-7R4As or 47,900 lb (21 747 kg) General Electric CF6-80As.

Performance: (Basic aircraft with JT9D-7R4Ds) Max. cruising speed, 582 mph (937 km/h) at 30,000 ft (9 145 m); long-range cruise, 528 mph (850 km/h) at 39,000 ft (11 890 m); max. payload range, 2,556 mls (4 114 km); max. fuel range, 5,642 mls (9 080 km).

Weights: Operational empty, 180,300 lb (81 856 kg); max. take-off, 300,000 lb (136 200 kg).

Accommodation: Flight crew of two plus optional third crew member on flight deck. Typical mixed-class seating for 18 six-abreast and 193 seven-abreast, with maximum single-class seating for 255 seven-abreast.

Status: Five aircraft to be used for preliminary certification with first scheduled to fly September 30, 1981, three more flying before year's end and customer deliveries commencing August 1982. Total of 166 (plus 135 on option) on order for 15 airlines at beginning of 1981.

Notes: The Model 767-100 is a shorter-body version, which, at the beginning of 1981, was considered unlikely to proceed as all orders to that time called for the longer -200.

BOEING 767-200

Dimensions: Span, 156 ft 4 in (47,65 m); length, 159 ft 2 in (48,50 m); height, 52 ft 0 in (15,85 m); wing area, 3,050 sq ft (283,3 m²).

BOEING E-3A SENTRY

Country of Origin: USA.
Type: Airborne warning and control system aircraft.
Power Plant: Four 21,000 lb (9 525 kg) Pratt & Whitney TF33-PW-100A turbofans.
Performance: (At max. weight) Average cruising speed, 479 mph (771 km/h) at 28,900–40,100 ft (8 810–12 220 m); average loiter speed, 376 mph (605 km/h) at 29,000 ft (8 840 m); time on station (unrefuelled) at 1,150 mls (1 850 km) from base, 6 hrs, (with one refuelling), 14·4 hrs; ferry range (crew reduced to four members), 5,034 mls (8 100 km) at 475 mph (764 km/h).
Weights: Empty, 170,277 lb (77 238 kg); normal loaded, 214,300 lb (97 206 kg); max. take-off, 325,000 lb (147 420 kg).
Accommodation: Operational crew of 17 comprising flight crew of four, systems maintenance team of four, a battle commander and an air defence operations team of eight.
Status: First of two (EC-137D) development aircraft flown February 9, 1972, two pre-production E-3As following in 1975. Twenty Sentries delivered by beginning of 1981 when last of 31 production examples for USAF scheduled for May 1984 delivery. First of 18 similar aircraft for NATO (excluding UK) scheduled for February 1981 completion, with last to be completed in June 1985.

BOEING E-3A SENTRY

Dimensions: Span, 145 ft 9 in (44,42 m); length, 152 ft 11 in (46,61 m); height, 42 ft 5 in (12,93 m); wing area, 2,892 sq ft (268,67 m²).

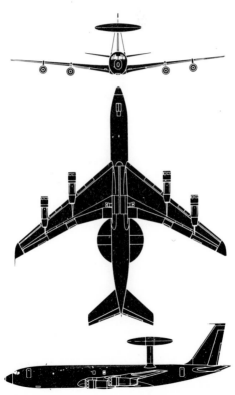

CANADAIR CL-600 CHALLENGER

Country of Origin: Canada.
Type: Light business executive transport.
Power Plant: Two 7,500 lb (3 405 kg) Avco Lycoming ALF 502L turbofans.
Performance: Max. speed, 562 mph (904 km/h); max. cruise, 553 mph (890 km/h); long-range cruise, 496 mph (800 km/h); range (effective from 60th aircraft), 3,910 mls (6 300 km) with (NBAA IFR) reserve of 230 mls/370 km; max. operating altitude, 49,000 ft (14 935 m).
Weights: (Effective from 60th aircraft) Empty, 18,450 lb (8 369 kg); operational empty, 22,675 lb (10 285 kg); max. take-off, 40,125 lb (18 201 kg).
Accommodation: Pilot and co-pilot side-by-side on flight deck and typical main cabin arrangement for 18 passengers, with optional layouts for 14, 15 and 17 passengers.
Status: First CL-600 (built on production jigs) flown on November 8, 1978, with second and third following on March 17 and July 14, 1979 respectively. First customer deliveries of CL-600 effected last quarter of 1980. First CL-601 scheduled to fly late 1981 with first customer deliveries late 1982, and first CL-610 scheduled to fly second quarter of 1982, with customer deliveries commencing mid-1983. Orders at beginning of 1981 comprised 130 CL-600s, 17 CL-601s and 35 CL-610s, with production to peak at seven monthly in 1981.
Notes: Three versions of the Challenger are in production or under development, the CL-600 (described and illustrated above), the CL-601 (illustrated opposite) with 9,140 lb (4 146 kg) General Electric CF34-1A turbofans, and the similarly-powered CL-610 Challenger E with 8 ft 9 in (2,67 m) fuselage "stretch" and extended (by 45 in/114 cm) wing.

CANADAIR CL-601 CHALLENGER

Dimensions: Span, 61 ft 10 in (18,85 m); length, 68 ft 5 in (20,85 m); height, 20 ft 8 in (6,30 m); wing area, 450 sq ft (41,82 m²).

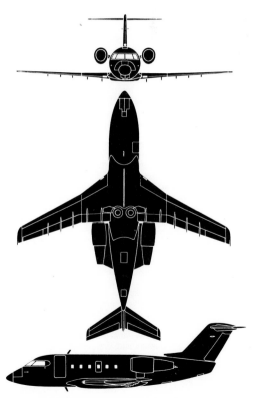

CAPRONI VIZZOLA C 22J

Country of Origin: Italy.

Type: Side-by-side two-seat primary-basic trainer.

Power Plant: Two 202 lb (92 kg) Microturbo TRS 18-046 turbojets, or (production option) two 242 lb (110 kg) Klöckner-Humboldt-Deutz KHD-317 turbojets.

Performance: (Estimated with KHD-317 engines) Max. speed, 329 mph (530 km/h) at 8,200 ft (2 500 m); max. continuous cruise, 292 mph (470 km/h) at 16,405 ft (5 000 m); econ. cruise, 186 mph (300 km/h) at 9,840 ft (3 000 m); initial climb, 2,070 ft/min (10,51 m/sec); time to 16,405 ft (5 000 m), 12 min; range (internal fuel with 10% reserves), 660 mls (1 060 km).

Weights: Empty, 1,124 lb (510 kg); normal load, 1,984 lb (900 kg); max. take-off, 2,425 lb (1 100 kg).

Armament: The C 22J may be fitted with two or four standard NATO underwing pylons, typical loads including four 97-lb (44-kg) or 440-lb (200-kg) practice bombs, two 7,62-mm gun pods with 500 rounds, or two pods each with 18 2-in (5-cm) rockets.

Status: Prototype C 22J flown July 21, 1980, and flight development continuing at beginning of 1981.

Notes: The C 22J is one of the first of a new generation of lightweight jet trainers (another being the Microjet 200—see pages 152–153) offering low initial procurement and minimum operational cost, and owing much to the manufacturer's earlier sailplane experience. The TRS 18 and KHD-317 turbojets are being offered as customer options. The fuselage pod of the C 22J, which utilises a fibreglass shell, is designed to act as a lifting body.

58

CAPRONI VIZZOLA C 22J

Dimensions: Span, 32 ft 9½ in (10,00 m); length, 20 ft 3¼ in (6,19 m); height, 6 ft 2 in (1,88 m); wing area, 94·19 sq ft (8,75 m²).

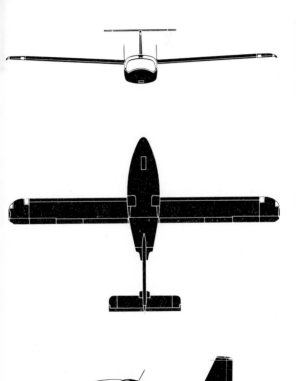

CASA C-101 AVIOJET

Country of Origin: Spain.

Type: Two-seat basic and advanced trainer.

Power Plant: One 3,500 lb (1 588 kg) Garrett AiResearch TFE 731-2-2J turbofan.

Performance: (At 10,362 lb/4 700 kg) Max. speed, 479 mph (770 km/h) or Mach 0·7 at 28,000 ft (8 535 m), 404 mph (650 km/h) or Mach 0·53 at sea level; time to 25,000 ft (7 620 m), 12 min; service ceiling, 41,000 ft (12 495 m); range (internal fuel at 11,540 lb/5 235 kg), 2,485 mls (4 000 km); max. climb (at 10,362 lb/4 700 kg), 3,350 ft/min (17 m/sec).

Weights: Basic operational empty, 6,790 lb (3 080 kg); loaded (pilot training mission with outer wing tanks empty), 10,362 lb (4 700 kg), (with max. internal fuel), 11,540 lb (5 235 kg); max. take-off, 12,346 lb (5 600 kg).

Armament: (C101ET) Seven external stores stations (six wing and one fuselage) for maximum of 3,307 lb (1 500 kg) of ordnance. Provision is made for a semi-recessed pod beneath the aft cockpit for a 30-mm cannon or two 7,62-mm Miniguns. Warload options include four Mk 83 or six Mk 82 bombs, or four AGM-65 Maverick missiles.

Status: Four prototypes of which first flown on June 29, 1977, and last on April 17, 1978. Sixty ordered by Spanish Air Force in March 1978, of which first four accepted by the Air Academy on March 17, 1980, and follow-on batch of additional 28 aircraft ordered shortly thereafter.

Notes: The standard C-101EB basic trainer is replacing the HA-200 Saeta, the T-6 Texan and the T-33A in the Spanish Air Force trainer inventory. No order has yet been placed for the armed C-101ET.

60

CASA C-101 AVIOJET

Dimensions: Span, 34 ft 9⅜ in (10,60 m); length, 40 ft 2¼ in (12,25 m); height, 13 ft 11 in (4,25 m); wing area, 215·3 sq ft (20,00 m²).

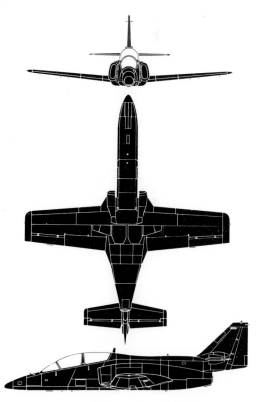

CASA C-212 AVIOCAR SERIES 200

Country of Origin: Spain.
Type: STOL utility transport and commuterliner.
Power Plant: Two 900 shp Garrett AiResearch TPE 331-10-501 C turboprops.
Performance: Max. cruising speed, 227 mph (365 km/h) at 10,000 ft (3 050 m), (at 80% power), 216 mph (347 km/h); range (max. payload and no reserves), 255 mls (410 km), (max. fuel and no reserves), 1,094 mls (1 760 km); max. climb rate, 1,555 ft/min (7,9 m/sec); service ceiling, 28,000 ft (8 535 m).
Weights: Max. take-off, 16,424 lb (7 450 kg).
Accommodation: Flight crew of two and (third-level airline arrangement) 26 passengers in four-abreast seating or 19 passengers in three-abreast seating.
Status: The first and second prototypes of the C-212 flew on March 26 and October 23, 1971 respectively, the first of eight pre-series aircraft following on November 17, 1972. The 138th and 139th production aircraft served as prototypes for the Series 200, flying on April 30 and June 20, 1978 respectively, the first production Series 200 (159th production aircraft to Dassault-Breguet) being delivered early 1980. Total sales of the C-212 (all versions) had reached 229 by the beginning of 1981 (including 68 for Indonesian assembly).
Notes: The Series 200 is a more powerful and structurally strengthened version of the Aviocar. An assembly line is operated in Indonesia by Nurtanio (as NC-212), the Series 200 Aviocar being introduced after completion of assembly of 28 of the initial model. The Aviocar serves the Spanish Air Force in transport, navigational trainer and photo survey versions.

CASA C-212 AVIOCAR SERIES 200

Dimensions: Span, 62 ft 4 in (19,00 m); length, 49 ft 10½ in (15,20 m); height, 20 ft 8¾ in (6,32 m); wing area, 430·56 sq ft (40,00 m²).

CESSNA MODEL 303 CRUSADER

Country of Origin: USA.

Type: Light cabin monoplane.

Power Plant: Two 250 hp Teledyne Continental TSIO-520-AE six-cylinder horizontally-opposed turbo-super-charged engines.

Performance: (Provisional) Max. speed, 247 mph (397 km/h) at 20,000 ft (6 095 m); max. continuous cruise, 207 mph (333 km/h) at 10,000 ft (3 050 m); range at max. cruise (max. fuel), 1,002 mls (1 612 km), (max. load), 575 mls (925 km); initial climb, 1,400 ft/min (7,1 m/sec); service ceiling, 25,000 ft (7 620 m).

Weights: Empty, 3,056 lb (1 386 kg); max. take-off, 5,025 lb (2 279 kg).

Accommodation: Six individual seats in three pairs with 10-in (25-cm) aisle, with baggage lockers in nose, wing and aft cabin for total of 590 lb (268 kg).

Status: First of two prototypes flown October 17, 1979, with first production deliveries schedules for August 1981.

Notes: Originally named Clipper, the Model 303 Crusader is claimed to offer a larger useful load than any comparable light twin. The two prototype Crusaders were scheduled to complete 300 and 200 hours trials plus 500 hours service testing by late 1980 prior to certification. Various internal arrangements are proposed, apart from the standard paired forward-seating described above, and including club seating with writing tables mounted on the sidewalls or the seat backs of forward-facing seats.

CESSNA MODEL 303 CRUSADER

Dimensions: Span, 38 ft 10 in (11,84 m); length, 30 ft 4⅞ in (9,26 m); height, 12 ft 10¾ in (3,93 m); wing area, 189 sq ft (17,56 m²).

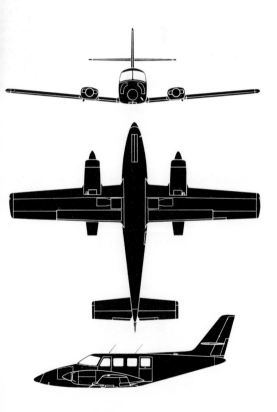

CESSNA MODEL 425 CORSAIR

Country of Origin: USA.

Type: Light business executive transport.

Power Plant: Two 450 shp Pratt & Whitney PT6A-112 turboprops.

Performance: Max. cruising speed (at 7,000 lb/3 175 kg), 304 mph (489 km/h) at 17,700 ft (5 395 m); range (at max. cruise), 1,620 mls (2 606 km) at 30,000 ft (9 145 m), 1,174 mls (1 890 km) at 20,000 ft (6 095 m); max. range, 1,895 mls (3 050 km) at 242 mph (389 km/h) at 30,000 ft (9 145 m), 1,577 mls (2 539 km) at 236 mph (380 km/h) at 20,000 ft (6 095 m); range with max. payload—1,762 lb/ 799 kg), 778 mls (1 252 km) at 293 mph (471 km/h); initial climb, 2,027 ft/min (10,3 m/sec).

Weights: Empty, 4,870 lb (2 209 kg); max. take-off, 8,200 lb (3 720 kg).

Accommodation: Pilot and up to seven passengers in individual paired seats with provision for 600 lb (272 kg) baggage in nose compartment and 500 lb (227 kg) in the aft cabin.

Status: Prototype Corsair flown in 1978, with eight delivered by beginning of 1981, when production was 10–12 monthly.

Notes: The Corsair is based on the airframe of the Model 421 Golden Eagle, which, with minor modifications, is mated with PT6A turboprops. The Corsair has a similar "wet" wing with integral fuel tankage.

CESSNA MODEL 425 CORSAIR

Dimensions: Span, 44 ft 1½ in (13,45 m); length, 35 ft 10¼ in (10,93 m); height, 12 ft 7⅓ in (3,84 m); wing area, 224·98 sq ft (20,90 m²).

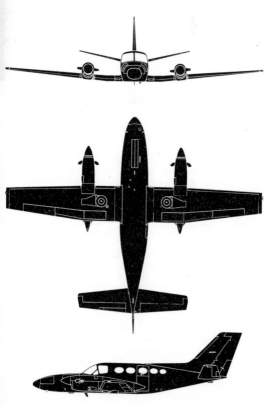

CESSNA CITATION II

Country of Origin: USA.

Type: Light business executive transport.

Power Plant: Two 2,500 lb (1 135 kg) Pratt & Whitney (Canada) JT15D-4 turbofans.

Performance: Max. cruise, 420 mph (676 km/h) at 25,400 ft (7 740 m); range cruise, 380 mph (611 km/h) at 43,000 ft (13 105 m); range (with eight passengers and 45 min reserve), 2,080 mls (3 347 km) at 380 mph (611 km/h); initial climb, 3,500 ft/min (17,8 m/sec); time to 41,000 ft (12 495 m), 34 min; max. cruise altitude, 43,000 ft (13 105 m).

Weights: Typical empty equipped, 6,960 lb (3 160 kg); max. take-off, 12,500 lb (5 675 kg).

Accommodation: Normal flight crew of two on separate flight deck and up to 10 passengers in main cabin.

Status: Two prototypes of Citation II flown January 31 and April 28, 1977 respectively, first customer deliveries commencing late March 1978, with 210 delivered by beginning of 1981, when combined production rate of Citation I and II was 16 monthly, 220 Citation Is having been delivered. These were preceded by 349 of original Citation.

Notes: The Citation II is a stretched (4 ft/1,22 m longer cabin) version of the original Citation, with a higher aspect ratio wing, uprated engines and increased fuel capacity, and is being manufactured in parallel with the Citation I and I/SP (the latter catering for single-pilot operation) with similar accommodation to the first Citation, JT15D-1A turbofans and a 47 ft 1 in/ 14,36 m wing. Citation I deliveries began in February 1977.

CESSNA CITATION II

Dimensions: Span, 51 ft 8 in (15,76 m); length, 47 ft 3 in (14,41 m); height 14 ft 11 in (4,55 m).

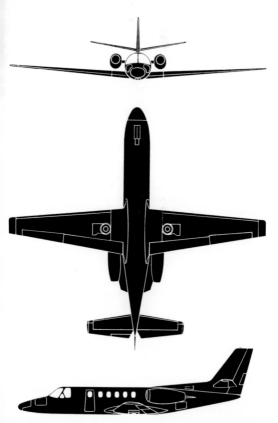

CESSNA CITATION III

Country of Origin: USA.

Type: Light business executive transport.

Power Plant: Two 3,650 lb (1 656 kg) Garrett AiResearch TFE 731-3B-100S turbofans.

Performance: Max. cruising speed, 540 mph (869 km/h) at 33,000 ft (10 060 m), 528 mph (850 km/h) at 41,000 ft (12 500 m), 509 mph (818 km/h) at 45,000 ft (13 715 m); range (two crew and four passengers), 2,735 mls (4 400 km) with 45 min reserves, (with six passengers and optional rear fuselage tank), 3,450 mls (5 555 km); max. initial climb, 4,475 ft/min (22,73 m/sec); ceiling, 51,000 ft (15 545 m).

Weights: Operational empty, 9,985 lb (4 529 kg); max. take-off, 19,500 lb (8 845 kg).

Accommodation: Normal flight crew of two on separate flight deck with standard cabin arrangement for six passengers in individual seats. Optional arrangements for eight to 13 passengers.

Status: First and second prototypes flown May 30, 1979 and May 2, 1980, with first customer deliveries scheduled for last quarter of 1982. Orders placed for some 150 Citation IIIs by beginning of 1981.

Notes: The Citation III possesses no commonality with the Citation II (see pages 68–69) despite its name, and is being offered in basic and extended-range versions, the latter augmenting the integral wing tankage with a rear fuselage tank. During the course of the prototype test programme, the ailerons and elevators have been redesigned, the wing trailing edge has been simplified, a one-piece main entry door has replaced the split airstair door, and the hydraulic and electrical systems have been revised.

CESSNA CITATION III

Dimensions: Span, 53 ft 3½ in (16,30 m); length, 55 ft 6 in (16,90 m); height, 17 ft 3½ in (5,30 m); wing area, 312 sq ft (29,00 m²).

DASSAULT-BREGUET ATLANTIC NG

Country of Origin: France.

Type: Long-range maritime patrol aircraft.

Power Plant: Two 6,105 ehp Rolls-Royce (SNECMA) Tyne RTy 20 Mk 21 turboprops.

Performance: (Estimated) Max. speed, 368 mph (593 km/h) at sea level, 409 mph (658 km/h) at 26,245 ft (8 000 m); max. cruise, 348 mph (560 km/h) at 19,685 ft (6 000 m); max. endurance, 16 hrs at 207 mph (333 km/h) at 19,685 ft (6 000 m).

Weights: Empty equipped, 55,280 lb (25 075 kg); max. loaded, 99,206 lb (45 000 kg); max. overload take-off, 101,852 lb (46 200 kg).

Armament: Homing torpedoes, mines, depth charges, etc, in ventral bay, plus four AM 39 ASMs on wing pylons.

Accommodation: Normal operating crew of 12, with flight crew of three on flight deck and remainder in nose, main tactical and rear compartments.

Status: First of two definitive prototypes to fly second quarter of 1981, and deliveries against French Navy requirement for 42 scheduled to commence first half of 1985 with completion by 1990.

Notes: The Atlantic NG (*Nouvelle Génération*) is a modernised version of the original Atlantic, production of which terminated in 1973 after completion of 87 series aircraft.

DASSAULT-BREGUET ATLANTIC NG

Dimensions: Span, 119 ft 1¼ in (36,30 m); length, 104 ft 1½ in (31,74 m); height, 33 ft 1¾ in (11,35 m); wing area, 1,291·7 sq ft (120 m²).

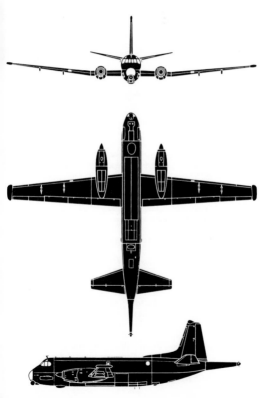

DASSAULT-BREGUET
MYSTÈRE-FALCON 20H

Country of Origin: France.

Type: Light business executive transport.

Power Plant: Two 5,050 lb (2 290 kg) Garrett AiResearch ATF3-6-1C turbofans.

Performance: Max. cruising speed, 528 mph (850 km/h) at 30,000 ft (9 145 m); range (with eight passengers and 45 min reserves), 2,703 mls (4 350 km) at 495 mph (795 km/h), 2,818 mls (4 535 km) at 460 mph (740 km/h) at 40,000 ft (12 190 m).

Weights: Empty equipped, 17,703 lb (8 030 kg); max. take-off, 32,000 lb (14 515 kg).

Accommodation: Crew of two on flight deck and normal seating for 8–10 passengers in main cabin, with high-density arrangement for 14 passengers.

Status: Intermediate prototype (Mystère-Falcon 20FH retaining CF 700 engines of 20F) flown April 24, 1979, with definitive prototype scheduled to fly February 1981, and certification scheduled for following May. Delivery of five (in Gardian maritime surveillance version) to French Navy scheduled from December 1982, with first commercial deliveries following in 1983.

Notes: The Mystère-Falcon 20H has been derived from the maritime surveillance 20G which (as the HU-25A Guardian) will enter service with the US Coast Guard mid-1981, 41 having been ordered with 5,300 lb (2 404 kg) ATF3-6-2C turbofans. The 20H has a similar fuel system to that of the Mystère-Falcon 50.

DASSAULT-BREGUET MYSTÈRE-FALCON 20H

Dimensions: Span, 53 ft 5¾ in (16,30 m); length, 56 ft 2⅞ in (17,14 m); height, 17 ft 5 in (5,32 m); wing area, 449·93 sq ft (41,80 m²).

DASSAULT-BREGUET
MYSTÈRE-FALCON 50

Country of Origin: France.

Type: Light business executive transport.

Power Plant: Three 3,700 lb (1 680 kg) Garrett AiResearch TFE731-3 turbofans.

Performance: Max. cruising speed, 547 mph (880 km/h) at 33,000 ft (10 060 m) or Mach 0·82; long-range cruise, 495 mph (792 km/h) at 37,000 ft (11 275 m) or Mach 0·75; range (with eight passengers and 45 min reserves), 4,088 mls (6 578 km) at long-range cruise, 2,764 mls (4 447 km) at max. cruise.

Weights: Empty equipped, 19,840 lb (9 000 kg); max. take-off, 38,800 lb (17 600 kg).

Accommodation: Flight crew of two and various cabin arrangements for 6–12 passengers.

Status: First and second prototypes flown on November 7, 1976 and February 18, 1978 respectively, with first pre-series aircraft flown June 13, 1978. Forty had been delivered by the beginning of 1981, when monthly production rate was four aircraft and orders exceeded 150.

Notes: Evolved from the Mystère-Falcon 20 series of business executive transports (see pages 74–75), the Mystère-Falcon 50 was modified subsequent to initial flight testing to incorporate a supercritical wing possessing the same planform as the original wing. The Series 20 and 50 Mystère-Falcons provide essentially similar passenger accommodation, but the latter has appreciably greater range and established a number of world records (C-1H class) during a non-stop ferry flight from Bordeaux to Teterboro on March 31, 1979. The example illustrated above serves with France's Ministerial Air Liaison Group.

DASSAULT-BREGUET MYSTÈRE-FALCON 50

Dimensions: Span, 61 ft 10½ in (18,86 m); length, 60 ft 9 in (18,52 m); height, 22 ft 10⅜ in (6,97 m); wing area, 504·13 sq ft (46,83 m²).

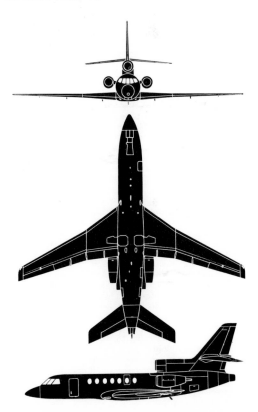

77

DASSAULT-BREGUET MIRAGE F1

Country of Origin: France.
Type: Single-seat multi-purpose fighter.
Power Plant: One 11,023 lb (5 000 kg) dry and 15,873 lb (7 200 kg) reheat SNECMA Atar 9K-50 turbojet.
Performance: Max. speed (clean), 915 mph (1 472 km/h) or Mach 1·2 at sea level, 1,450 mph (2 335 km/h) or Mach 2·2 at 39,370 ft (12 000 m); range cruise, 550 mph (885 km/h) at 29,530 ft (9 000 m); range with max. external fuel, 2,050 mls (3 300 km), with max. external combat load of 8,818 lb (4 000 kg), 560 mls (900 km), with external combat load of 4,410 lb (2 000 kg), 1,430 mls (2 300 km); service ceiling, 65,600 ft (20 000 m).
Weights: Empty, 16,314 lb (7 400 kg); loaded (clean), 24,030 lb (10 900 kg); max. take-off, 32,850 lb (14 900 kg).
Armament: Two 30-mm DEFA cannon and (intercept) 1-3 Matra 530 Magic and two AIM-9 Sidewinder AAMs.
Status: First of four prototypes flown December 23, 1966. First production for *Armée de l'Air* flown February 15, 1973. Production rate of seven per month at beginning of 1981. Licence manufacture is being undertaken in South Africa. Firm orders totalled 649 aircraft by beginning of 1981 including Greece, 40 (F1CG), Kuwait, 20 (18 F1CK and two F1BK), Libya, 38 (32 F1ED and six F1BD), Iraq, 60 (56 F1EQ and four F1BQ), Jordan, 17, Morocco, 50 (F1CH), South Africa, 48 (16 F1CZ and 32 F1AZ), Spain, 73 (six F1BE, 45 F1CE and 22 F1EE), Qatar, 14 and Ecuador, 18. The *Armée de l'Air* plans total procurement of 246.
Notes: Production versions currently comprise F1A and F1E for ground attack role, the former for VFR operations only, the F1BD tandem two-seat conversion trainer (illustrated above), 14 of which have been ordered by the *Armée de l'Air*, and the F1C interceptor.

DASSAULT-BREGUET MIRAGE F1

Dimensions: Span, 27 ft 6¾ in (8,40 m); length, 49 ft 2½ in (15,00 m); height, 14 ft 9 in (4,50 m); wing area, 269·098 sq ft (25 m²).

DASSAULT-BREGUET MIRAGE 2000

Country of Origin: France.
Type: Single-seat multi-role fighter.
Power Plant: One 12,230 lb (5 600 kg) dry and 19,840 lb (9 000 kg) reheat SNECMA M53-5 turbofan.
Performance: Max. speed (clean), 1,550 mph (2 495 km/h) above 36,090 ft (11 000 m) or Mach 2·35, 915 mph (1 472 km/h) at sea level or Mach 1·2; tactical radius (intercept mission with four AAMs and two 374 Imp gal/1 700 l drop tanks), 435 mls (700 km); max. climb, 49,000 ft/min (249 m/sec); time to Mach 2·0 at 49,200 ft (15 000 m) from brakes release, 4·0 min.
Weights: Combat, 19,840 lb (9 000 kg); max. take-off, 33,070 lb (15 000 kg).
Armament: Two 30-mm DEFA 554 cannon and (air superiority) two Matra 550 Magic and two Matra Super 530D AAMs, or (strike) up to 11,000 lb (5 000 kg) of ordnance on nine external stations.
Status: First prototype flown March 10, 1978, and fifth and last prototype (two-seat Mirage 2000B) on October 11, 1980. Total of 48 ordered (against total *Armée de l'Air* requirement for some 400 in three main versions) by beginning of 1981, with deliveries commencing 1983.
Notes: Three major versions—single-seat interceptor and attack, and two-seat tactical nuclear strike—are programmed for the *Armée de l'Air*, and from 1985, production aircraft will be powered by the M53-P2 with a military rating of 14,330 lb (6 500 kg) and 21,385 lb (9 700 kg) with maximum reheat. Deliveries of the Mirage 2000 ASMP two-seat nuclear strike aircraft are scheduled to start in 1986.

DASSAULT-BREGUET MIRAGE 2000

Dimensions: Span, 29 ft 6¼ in (9,00 m); length, 50 ft 3½ in (15,33 m); wing area, 441·3 sq ft (41,00 m²).

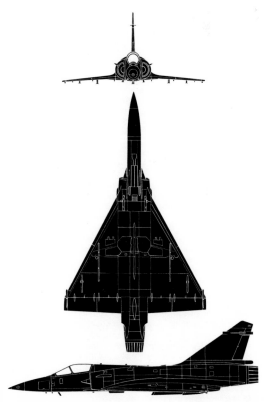

DASSAULT-BREGUET SUPER MIRAGE 4000

Country of Origin: France.

Type: Single-seat multi-role fighter.

Power Plant: Two 12,230 lb (5 600 kg) dry and 19,840 lb (9 000 kg) reheat SNECMA M53-5 turbofans.

Performance: (Estimated) Max. sustained speed, 1,452 mph (2 336 km/h) or Mach 2·2 above 36,090 ft (11 000 m), 915 mph (1 472 km/h) or Mach 1·2 at sea level; max. climb rate, 50,000 ft/min (254 m/sec); operational ceiling, 65,000 ft (19 810 m).

Weights: (Estimated) Loaded (clean), 37,500 lb (17 000 kg); max. take-off, 45,000 lb (20 410 kg).

Armament: Two 30-mm DEFA 554 cannon and up to 15,000 lb (6 804 kg) of ordnance on nine external stations (four wing and five fuselage).

Status: Sole prototype Super Mirage 4000 flown on March 9, 1979. Flight and systems development continuing at the beginning of 1981.

Notes: Developed as a private venture, the Super Mirage 4000 is optimised for the deep penetration role but is also suitable for intercept and air superiority missions, and is most closely comparable with the McDonnell Douglas F-15 Eagle. It closely resembles the Mirage 2000 (see pages 80–81) in aerodynamic, structural and systems layout, sharing with the smaller aircraft such features as fly-by-wire controls, artificial stability, leading-edge flaps and the use of carbon-fibre composites, components using these materials including the fin and rudder, the elevons and the canard surfaces attached to the outer sides of the intake ducts. The prototype is to be equipped with the Thomson-CSF RDM multi-mode radar and other weapon system equipment from the Mirage 2000.

82

DASSAULT-BREGUET SUPER MIRAGE 4000

Dimensions: Span, 39 ft 4½ in (12,00 m); length, 61 ft 4¼ in (18,70 m); wing area, 786 sq ft (73,00 m²).

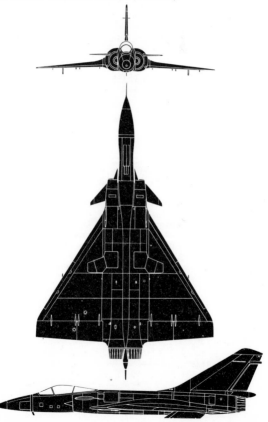

DASSAULT-BREGUET/DORNIER ALPHA JET

Countries of Origin: France and Federal Germany.
Type: Two-seat basic-advanced trainer and light tactical aircraft.
Power Plant: Two 2,975 lb (1 350 kg) SNECMA-Turboméca Larzac 04-C5 turbofans.
Performance: Max. speed, 622 mph (1 000 km/h) at sea level or Mach 0·816, 567 mph (912 km/h) at 32,810 ft (10 000 m) or Mach 0·84; tactical radius (training mission LO-LO-LO profile), 267 mls (430 km); ferry range (max. internal fuel), 1,243 mls (2 000 km), (with two 68 Imp gal/310 l external tanks), 1,678 mls (2 700 km); max. climb, 11,220 ft/min (57 m/sec); ceiling, 45,000 ft (13 715 m).
Weights: Empty, 7,716 lb (3 500 kg); normal loaded (clean), 11,023 lb (5 000 kg); normal take-off (close air support), 13,448 lb (6 100 kg); max. overload, 15,983 lb (7 250 kg).
Armament: External centreline gun pod with (Alpha Jet E) 30-mm DEFA 533 or (Alpha Jet A) 27-mm Mauser cannon. Max. load of 4,850 lb (2 200 kg) on external stations.
Status: First of four prototypes flown October 26, 1973, with first production Alpha Jet E flying on November 4, 1977, and first production Alpha Jet A on April 12, 1978. Combined production rate of 12 monthly from French and German lines at beginning of 1981, with 200th delivered November 1980 against orders for 175 for France, 175 for Germany, 33 for Belgium, six for Ivory Coast, 24 for Morocco, 12 for Nigeria, six for Qatar and five for Togo.
Notes: Final assembly lines in Toulouse and Munich. The *Armée de l'Air* version is optimised for basic-advanced training and the *Luftwaffe* model (illustrated by photo) is a dedicated close air support aircraft.

DASSAULT-BREGUET/DORNIER ALPHA JET

Dimensions: Span, 29 ft 11 in (9,11 m); length, 40 ft 3 in (12,29 m); height, 13 ft 9 in (4,19 m); wing area, 188 sq ft (15,50 m²).

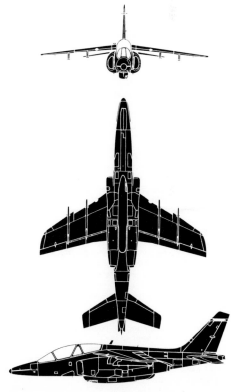

DE HAVILLAND CANADA DASH 7-100

Country of Origin: Canada.
Type: STOL short-haul commercial transport.
Power Plant: Four 1,120 shp Pratt & Whitney PT6A-50 turboprops.
Performance: Max. cruising speed, 266 mph (428 km/h) at 8,000 ft (2 440 m), 262 mph (421 km/h) at 15,000 ft (4 570 m); long-range cruise, 248 mph (399 km/h); range (with 50 passengers and IFR reserves), 795 mls (1 280 km); max. fuel range (with 6,500-lb/2 948-kg payload), 1,347 mls (2 168 km).
Weights: Empty equipped (typical), 27,600 lb (12 519 kg); max. take-off, 44,000 lb (19 958 kg).
Accommodation: Flight crew of two and standard seating for 50 passengers four-abreast.
Status: Two pre-production aircraft flown March 27 and June 26, 1975, and first production aircraft following on May 30, 1977, with first customer delivery (to Rocky Mountain) October 1977. Orders totalled 65 (plus 54 on option) at beginning of 1981, when production rate was three monthly with some 30 delivered.
Notes: Developed versions of the basic Dash 7 will include the Series 150 with a higher gross weight, the Series 200 with 1,230 shp PT6A-55 engines and the Series 300 with a 70-in (178-cm) fuselage stretch increasing passenger capacity to 60. A maritime surveillance variant, the Dash 7R Ranger, has been ordered for the Canadian Coast Guard. This offers extended , payload and range performance, and increased fuel capacity, and the two examples covered by initial CCG procurement are to enter sevice in 1981.

DE HAVILLAND CANADA DASH 7-100

Dimensions: Span, 93 ft 0 in (28,35 m); length, 80 ft 7¾ in (24,58 m); height, 26 ft 2 in (7,98 m); wing area, 860 sq ft (79.90 m²).

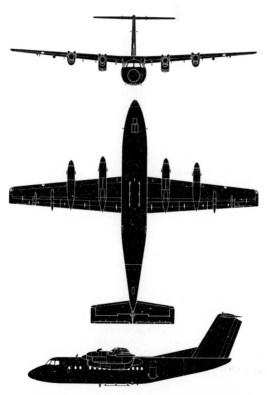

DORNIER DO 128-6

Country of Origin: Federal Germany.

Type: Light utility transport.

Power Plant: Two 400 shp Pratt & Whitney PT6A-110 turboprops (derated from 475 shp).

Performance: Max. speed, 211 mph (340 km/h); max. range cruise, 159 mph (256 km/h); max. fuel range, 1,030 mls (1 658 km); max. climb, 1,260 ft/min (6,4 m/sec); service ceiling, 28,150 ft (8 580 m).

Weights: Operational empty, 5,600 lb (2 540 kg); max. take-off, 9,480 lb (4 300 kg).

Accommodation: Pilot and either co-pilot or passenger on flight deck and standard accommodation for nine passengers with aisle in commuterliner arrangement. Cabin may be stripped for freight carrying, and 13 inward-facing folding seats may be fitted.

Status: The prototype Do 128-6 flew on March 4, 1980, and deliveries of this type were scheduled to commence in the first quarter of 1981.

Notes: The Do 128 is an upgraded version of the Do 28D-2 Skyservant, production of which (with 380 hp Avco Lycoming IGSO-540A1E piston engines) is continuing as the Do 128-2 which has increased take-off and landing weights, and a higher payload. Apart from engines, the Do 128-6 (formerly known as the Do 28D-6) and the Do 128-2 are similar, and the latter was being offered at the beginning of 1981 for maritime surveillance with MEL Marec radar. The original Do 28D prototype was first flown on February 23, 1966, and the initial turboprop-powered model, the Do 28D-5X with Lycoming LTP 101-600 engines (see 1979 edition) was flown on April 9, 1978.

88

DORNIER DO 128-6

Dimensions: Span, 51 ft 1 in (15,55 m); length, 37 ft 5 in (11,41 m); height, 12 ft 9½ in (3,90 m); wing area, 312·2 sq ft (29,00 m²).

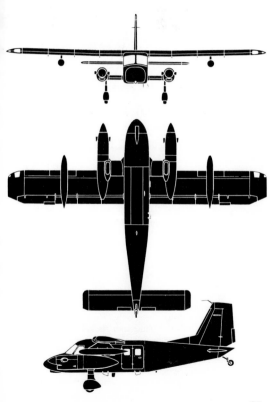

DORNIER DO 228

Country of Origin: Federal Germany.
Type: Light commuterliner and utility transport.
Power Plant: Two 715 shp Garrett AiResearch TPE 331-5 turboprops.
Performance: (Estimated for Do 228-100) Max. speed, 268 mph (432 km/h) at 10,000 ft (3 280 m), 230 mph (370 km/h) at sea level; max. range cruise, 206 mph (332 km/h) at 10,000 ft (3 280 m); range (with max. passenger payload), 1,075 mls (1 730 km); max. climb rate, 2,050 ft/min (10,4 m/sec); service ceiling, 29,600 ft (9 020 m).
Weights: Operational empty (-100), 7,040 lb (3 193 kg), (-200), 7,370 lb (3 343 kg); max. take-off (both versions), 12,570 lb (5 700 kg).
Accommodation: Flight crew of two with (standard passenger versions) 15 (-100) or 19 (-200) individual seats with central aisle.
Status: Prototypes of both -100 and -200 versions of the Do 228 are scheduled to fly spring 1981, with customer deliveries commencing early 1982.
Notes: The Do 228-100 (illustrated above) and -200 (illustrated opposite) differ in overall length, that of the -100 being 49 ft 3 in (15,03 m). The Do 228 mates the fuselage cross section of the Do 128 with a new-technology wing of supercritical section and utilising glassfibre composites for the raked wingtips and leading edge, and carbonfibre composites for the trailing edge movable surfaces.

DORNIER DO 228-200

Dimensions: Span, 55 ft 7 in (16,97 m); length, 54 ft 3 in (16,55 m); height, 15 ft 9 in (4,86 m); wing area, 344·46 sq ft (32,00 m²).

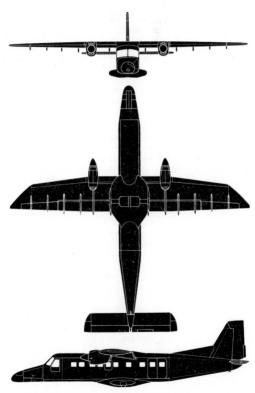

EDGLEY EA7 OPTICA

Country of Origin: United Kingdom.
Type: Three-seat observation aircraft.
Power Plant: One 160 hp Avco Lycoming O-320-B2B four-cylinder horizontally-opposed engine driving ducted fan.
Performance: Max. speed, 126 mph (203 km/h); cruise (65% power), 108 mph (174 km/h); range (at 65% power), 620 mls (1 000 km); loiter endurance, 13 hrs; max. initial climb, 720 ft/min (3,66 m/sec); service ceiling, 8,000 ft (2 440 m).
Weights: Empty, 1,973 lb (895 kg); max. take-off, 2,690 lb (1 220 kg).
Status: Prototype Optica flown on December 14, 1979, and negotiations under way at beginning of 1981 for the sub-contract manufacture of an initial batch of 26 aircraft.
Notes: Intended for pipeline and powerline inspection, traffic surveillance, forestry, coastal and frontier patrol, and aerial photography, the Optica is of extremely unusual design and is intended to afford the best possible all-round view that can be achieved with a fixed-wing aircraft. The O-320-B2B engine is part of a ducted propulsor unit which forms a power pod separate from the main shroud and is mounted downstream of a five-bladed fixed-pitch fan. The considerable flap area confers STOL capability on the Optica, a low wing loading, pre-set flaps and a low stalling speed permitting continuous en-route flight at very low speeds.

EDGLEY EA7 OPTICA

Dimensions: Span, 39 ft 10¾ in (12,16 m); length, 26 ft 9¼ in (8,16 m); height, 6 ft 3½ in (1,92 m); wing area, 170·5 sq ft (15,84 m²).

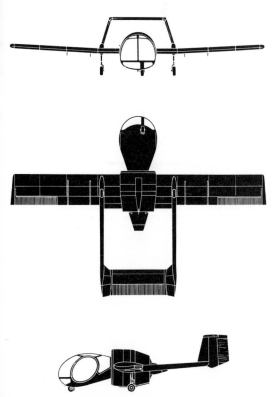

EMBRAER EMB-110P2 BANDEIRANTE

Country of Origin: Brazil.

Type: Third-level commuter transport.

Power Plant: Two 750 shp Pratt & Whitney PT6A-34 turboprops.

Performance: Max. speed, 286 mph (460 km/h) at 8,000 ft (2 440 m); max. cruise, 259 mph (417 km/h) at 10,000 ft (3 050 m); econ. cruise, 203 mph (326 km/h); range (max. fuel and 45 min reserves), 1,180 mls (1 900 km), (with 3,175-lb/1 440-kg payload), 309 mls (497 km).

Weights: Empty equipped, 7,751 lb (3 515 kg); max. take-off, 12,566 lb (5 700 kg).

Accommodation: Pilot and co-pilot side-by-side on flight deck and up to 21 passengers in seven rows three abreast in main cabin.

Status: The first EMB-110P2 (146th Bandeirante) flown May 3, 1977, with deliveries commencing in following year. Original prototype Bandeirante flown on October 26, 1968, and approximately 310 Bandeirantes of all versions (excluding EMB-111) delivered by beginning of 1981 (including 110 for Brazilian Air Force) when production tempo was 6–7 monthly.

Notes: The Bandeirante (Pioneer) has been the subject of continuous development for the past 12 years, the principal current versions being the EMB-110P1 and P2, the latter, described and illustrated, lacking the enlarged cargo door of the former (which has quick-change seating for 18 passengers). The EMB-110K1 is an all-cargo equivalent of the P1, and the EMB-110P3, scheduled to fly in 1982, is a pressurised version for 19 passengers and will feature a T-type tail, uprated PT6A-65 engines, slightly extended wingtips and a modified undercarriage.

EMBRAER EMB-110P2 BANDEIRANTE

Dimensions: Span, 50 ft 3⅛ in (15,32 m); length, 49 ft 5¾ in (15,08 m); height, 16 ft 1¾ in (4,92 m); wing area, 312 sq ft (29,00 m²).

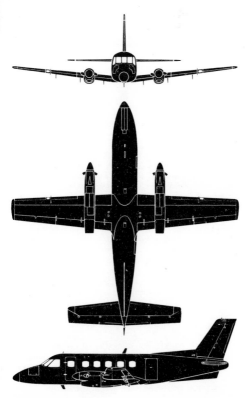

95

EMBRAER EMB-111

Country of Origin: Brazil.
Type: Maritime patrol and coastal surveillance aircraft.
Power Plant: Two 750 shp Pratt & Whitney (Canada) PT6A-34 turboprops.
Performance: Max. cruising speed, 239 mph (385 km/h) at 9,840 ft (3 000 m); econ. cruise, 219 mph (352 km/h) at 10,000 ft (3 050 m); average patrol speed, 198 mph (318 km/h) at 2,000 ft (610 m); range (max. fuel and 45 min reserves), 1,830 mls (2 945 km) at 10,000 ft (3 050 m); max. initial climb, 1,190 ft/min (6,04 m/sec); service ceiling (at 11,684 lb/5 300 kg), 25,500 ft (7 770 m).
Weights: Empty equipped, 8,289 lb (3 760 kg); max. take-off, 15,432 lb (7 000 kg).
Armament: Four underwing pylons for 127-mm air-to-surface rockets (two per pylon), or three pylons plus a leading-edge mounted 50 million candlepower searchlight.
Accommodation: Pilot and co-pilot side-by-side, and navigator, observer and radio/radar operator in main cabin.
Status: Twelve EMB-111 Ms ordered by Brazilian Air Force, these including two prototypes, the first of which flew on August 15, 1977. Six EMB-111Ns have been ordered by the Chilean Navy, deliveries of these being completed in 1979, and one EMB-111 has been ordered by the Gabon Air Force for 1981 delivery.
Notes: The EMB-111 is a derivative of the EMB-110 Bandeirante (see pages 94—95), the principal external differences being a nose radome and wingtip fuel tanks. The Chilean aircraft have some mission equipment changes, including passive ECM antennae under the nose and at the tail.

EMBRAER EMB-111

Dimensions: Span, 52 ft 4¾ in (15,95 m); length, 48 ft 11 in (14,91 m); height, 15 ft 10 in (4,83 m); wing area, 313·24 sq ft (29,10 m²).

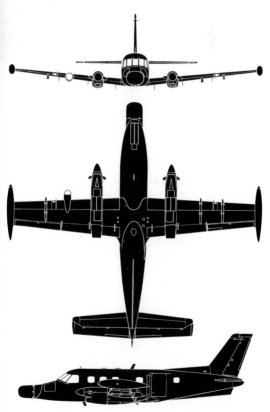

EMBRAER EMB-121 XINGU

Country of Origin: Brazil.

Type: Light business executive transport.

Power Plant: Two 680 shp Pratt & Whitney PT6A-28 turboprops.

Performance: Max. cruise, 280 mph (450 km/h) at 11,000 ft (3 353 m); initial climb, 1,400 ft/min (7,11 m/sec); service ceiling, 26,000 ft (7 925 m); range (with 1,985-lb/900-kg payload), 1,036 mls (1 668 km) at 20,000 ft (6 100 m); max. range (with 1,344-lb/610-kg payload), 1,462 mls (2 353 km).

Weights: Empty equipped, 7,716 lb (3 500 kg); max. take-off, 12,500 lb (5 670 kg).

Accommodation: Two seats side-by-side on flight deck and individual seats for five–six passengers in main cabin.

Status: Prototype flown October 10, 1976, with first production aircraft on May 20, 1977. Some 44 delivered by beginning of 1981 when production rate was two monthly. Forty-one ordered by French government (25 for *Armée de l'Air* and 16 for *Aéronavale*) with eight to be delivered by end of 1981 and final 14 in 1983.

Notes: A "stretched" version, the Xingu 2, with a 33-in (84-cm) increase in fuselage length raising passenger capacity to eight and uprated PT6A-42 engines was under development at the beginning of 1981.

EMBRAER EMB-121 XINGU

Dimensions: Span, 47 ft 5 in (14,45 m); length, 40 ft 2¼ in (12,25 m); height, 15 ft 6½ in (4,74 m); wing area, 296 sq ft (27,50 m²).

EMBRAER EMB-312 (T-27)

Country of Origin: Brazil.
Type: Tandem two-seat basic trainer and light strike aircraft.
Power Plant: One 750 shp Pratt & Whitney PT6A-25C turboprop.
Performance: Max. speed (at 3,307 lb/1 500 kg), 302 mph (486 km/h) at 10,000 ft (3 050 m), (at 5,181 lb/2 350 kg), 296 mph (476 km/h) at 8,000 ft (2 440 m); max. continuous cruise, 272 mph (438 km/h); initial climb (at 5,070 lb/2 300 kg), 2,126 ft/min (10,80 m/sec); max. range (30 min reserves), 1,312 mls (2 112 km); tactical radius HI-LO-HI (with four 250-lb/113-kg Mk 81 bombs), 161 mls (260 km), (with two Mk 81 bombs), 602 mls (970 km).
Weights: Max. take-off, 5,180 lb (2 350 kg).
Armament: Two 0·5-in (12,7-mm) machine gun pods with 350 rpg, four pods each with seven 37-mm or 70-mm rockets, or four 250-lb (113-kg) Mk 81 bombs.
Status: First of four prototypes flown on August 16, 1980, with deliveries against order for 168 for the Brazilian Air Force scheduled to commence second half of 1982 at a rate of five monthly.
Notes: Unique among current turboprop-driven basic trainers in having ejection seats for pupil and instructor, the EMB-312 has been designed to a specification drafted by the Brazilian Air Force by which it is designated T-27.

EMBRAER EMB-312 (T-27)

Dimensions: Span, 36 ft 6½ in (11,14 m); length, 32 ft 4¼ in (9,86 m); height, 11 ft 1⅞ in (3,40 m); wing area, 204·52 sq ft (19,00 m²).

FAIRCHILD A-10A THUNDERBOLT II

Country of Origin: USA.

Type: Single-seat close-support aircraft.

Power Plant: Two 9,065 lb (4 112 kg) General Electric TF34-GE-100 turbofans.

Performance: (At 38,136 lb/17 299 kg) Max. speed, 433 mph (697 km/h) at sea level, 448 mph (721 km/h) at 10,000 ft (3 050 m); initial climb, 5,340 ft/min (27,12 m/sec); service ceiling, 34,700 ft (10 575 m); combat radius (with 9,540-lb/4 327-kg bomb load and 1,170 lb/531 kg of 30-mm ammunition, with allowance for 1·93 hr loiter at 5,000 ft/1 525 m), 288 mls (463 km) at (average) 329 mph (529 km/h) at 25,000–35,000 ft (7 620–10 670 m); ferry range, 2,487 mls (4 000 km).

Weights: Empty, 19,856 lb (9 006 kg); basic operational, 22,844 lb (10 362 kg); max. take-off, 46,786 lb (22 221 kg).

Armament: One seven-barrel 30-mm General Electric GAU-8 Avenger rotary cannon. Eleven external stations for maximum of 9,540 lb (4 327 kg) ordnance (with full internal fuel and 1,170 lb/531 kg ammunition), or max. of 16,000 lb (7 250 kg).

Status: First of two prototypes flown May 10, 1972, and first of six pre-production aircraft flown February 15, 1975. First production aircraft flown October 21, 1975, and 400th delivered September 1980, with 825 planned for delivery by April 1986 (for USAF, Air National Guard and Air Force Reserve).

Notes: Planning at beginning of 1981 called for 30 (half Fiscal 1981 procurement) A-10s to be completed as tandem two-seat combat-ready trainers, the two-seat airframe being based on that of the experimental night and adverse weather version (see 1980 edition), which, flown on May 4, 1979, has 94% structural commonality with the single-seater.

FAIRCHILD A-10A THUNDERBOLT II

Dimensions: Span, 57 ft 6 in (17,53 m); length, 53 ft 4 in (16,25 m); height, 14 ft 8 in (4,47 m); wing area, 506 sq ft (47,01 m²).

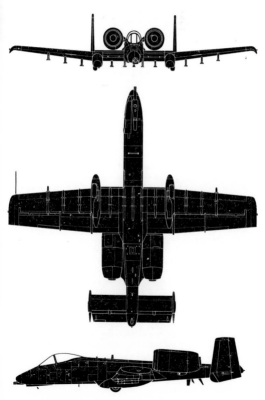

FOKKER F27 MARITIME

Country of Origin: Netherlands.
Type: Medium-range maritime surveillance aircraft.
Power Plant: Two 2,370 eshp Rolls-Royce Dart 536-7R turboprops.
Performance: Max. speed, 294 mph (474 km/h) at 20,000 ft (6 100 m); cruise (at 38,000 lb/17 235 kg), 288 mph (463 km/h) at 20,000 ft (6 100 m); typical search speed, 167–201 mph (269–324 km/h); max. range (with 30 min loiter and 5% fuel reserve), 3,107 mls (5 000 km); (transport mission with 10,000-lb/4 540-kg payload), 932 mls (1 500 km).
Weights: Operational equipped, 29,352 lb (13 314 kg); normal loaded, 45,000 lb (20 412 kg); max. take-off (overload), 47,500 lb (21 546 kg).
Accommodation: Flight crew of two with provision on flight deck for observer or engineer, and typical maritime surveillance crew complement of two observers, radar operator and tactical commander/navigator.
Status: Prototype Maritime (converted F27 Mk 100) flown February 28, 1976, and first production examples (for the Peruvian Navy) flown June 14 and September 28, 1977. Eleven ordered by beginning of 1981, for Peruvian Navy (2), Spanish Air Force (3), Angolan Air Force (1), Philippine Air Force (3), and Netherlands Navy (2), with six delivered.
Notes: The Maritime is based on either the Friendship Mk 200 or Mk 400 (forward freight door and reinforced floor) airframe, with search radar, long-range inertial navigation, blister windows, pylon tank provision, etc. F27 orders (all versions) totalled 517 (excluding 205 built by Fairchild) at the beginning of 1981.

FOKKER F27 MARITIME

Dimensions: Span, 95 ft 1⅘ in (29,00 m); length, 77 ft 3½ in (23,56 m); height, 28 ft 6⁷⁄₁₀ in (8,70 m); wing area, 753·47 sq ft (70,00 m²).

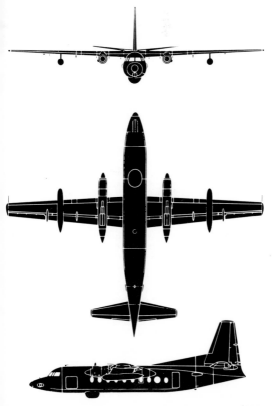

FOKKER F28 FELLOWSHIP Mᴋ 4000

Country of Origin: Netherlands.
Type: Short-haul commercial transport.
Power Plant: Two 9,850 lb (4 468 kg) Rolls-Royce RB. 183-2 Spey Mk 555-15H turbofans.
Performance: Max. cruise, 523 mph (843 km/h) at 23,000 ft (7 000 m); econ. cruise, 487 mph (783 km/h) at 32,000 ft (9 755 m); range cruise, 421 mph (678 km/h) at 30,000 ft (9 145 m); range (with max. payload), 1,160 mls (1 870 km) at econ. cruise, (with max. fuel), 2,566 mls (4 130 km); max. cruise altitude, 35,000 ft (10 675 m).
Weights: Operational empty (typical), 37,736 lb (17 117 kg); max. take-off, 71,000 lb (32 200 kg).
Accommodation: Flight crew of two and typical single-class configuration for 85 passengers five abreast.
Status: First and second F28 prototypes flown May 9 and August 3, 1967, first delivery following on February 24, 1969. A total of 169 F28s (all versions) ordered by 1981, when production rate was 1·25 monthly.
Notes: The F28 Mks 1000 and 2000 are now out of production (after completion of 97 and 10 respectively), having been replaced by the Mks 3000 and 4000, both having unslatted, longer-span wings and Spey Mk 555-15H engines. The former has the 80 ft 6½ in (24,55 m) fuselage of the Mk 1000 and the latter has the longer fuselage of the Mk 2000. Also on offer is the slatted Mk 6000 (see 1977 edition) with the same high-density accommodation as the Mk 4000.

106

FOKKER F28 FELLOWSHIP Mk 4000

Dimensions: Span, 82 ft 3 in (25,07 m); length, 97 ft 1¾ in (29,61 m); height, 27 ft 9½ in (8,47 m); wing area, 850 sq ft (78,97 m²).

GATES LEARJET 35A

Country of Origin: USA.

Type: Light business executive transport.

Power Plant: Two 3,500 lb (1 588 kg) Garrett AiResearch TFE 731-2-2B turbofans.

Performance: Max. speed, 542 mph (872 km/h) at 25,000 ft (7 620 m); max. cruise, 528 mph (850 km/h) at 41,000 ft (12 500 m); econ. cruise, 481 mph (774 km/h) at 45,000 ft (13 715 m); range (four passengers and max. fuel with 45 min reserves), 2,789 mls (4 488 km); max. initial climb, 4,900 ft/min (24,89 m/sec); service ceiling, 45,000 ft (13 715 m).

Weights: Empty equipped, 9,271 lb (4 205 kg); max. take-off, 17,000 lb (7 711 kg).

Accommodation: Crew of two on flight deck and up to eight passengers in main cabin.

Status: Learjet 35 prototype flown August 22, 1973 (as a turbofan-powered growth version of the Learjet 25), customer deliveries commencing November 1974. The improved Learjet 35A appeared in 1976, and, with progressive refinement, has since been in continuous production (in parallel with the Learjet 36A). By the beginning of 1981, some 380 of the turbofan-powered Models 35(A)/36(A) had been delivered and production was rising from six to seven/eight monthly.

Notes: The Learjet 35A and 36A are almost identical, differing in fuel capacity and accommodation, the latter having passenger accommodation reduced to six persons and provision for an extra fuel tank at the rear of the cabin. The 1980 Learjet 35A introduced an additional window on each side of the cabin, and Sea Patrol versions of both the Learjet 35A and 36A (see 1980 edition) have been developed. Three multi-role military Learjet 35As are to be delivered to the Finnish Air Force in 1982 for maritime patrol, target towing, etc.

108

GATES LEARJET 35A

Dimensions: Span, 39 ft 6 in (12,04 m); length, 48 ft 8 in (14,33 m); height, 12 ft 3 in (3,73 m); wing area, 253·3 sq ft (23,5 m²).

GATES LEARJET LONGHORN 55

Country of Origin: USA.
Type: Light business executive transport.
Power Plant: Two 3,700 lb (1 678 kg) Garrett AiResearch TFE 731-3-2B turbofans.
Performance: Max. speed, 534 mph (860 km/h) at 30,000 ft (9 150 m); max. cruise, 506 mph (814 km/h) at 45,000 ft (13 715 m); econ. cruise, 462 mph (743 km/h) at 49,000 ft (14 935 m); range (with max. fuel, four passengers and 45 min reserves), 2,859 mls (4 600 km); max. initial climb, 4,675 ft/min (23,75 m/sec); service ceiling, 51,000 ft (15 545 m).
Weights: Empty equipped, 10,992 lb (4 986 kg); max. take-off, 19,500 lb (8 845 kg).
Accommodation: Flight crew of two and up to 11 passengers in main cabin.
Status: Two Longhorn 55 prototypes flown on April 19 and November 15, 1979, and first production aircraft following on August 11, 1980, with customer deliveries scheduled to commence April–May 1981. Orders for the Longhorn 55 (and 56) exceeded 150 by beginning of 1981, when production tempo was building up to five monthly.
Notes: Combining the wing development of the Longhorn 28/29 (see 1979 edition) with an entirely new fuselage, the Longhorn 55 and 56 are externally identical, but the latter has increased fuel capacity and a higher max. take-off weight, range being 3,466 miles (5 578 km) and accommodation being provided for eight passengers. The Longhorn 55 and 56 supplement the existing range of (smaller) Learjets, and combined production of all versions is scheduled to be running at 17 per month by early 1982.

GATES LEARJET LONGHORN 55

Dimensions: Span, 43 ft 9½ in (13,34 m); length, 55 ft 1½ in (16,79 m); height, 14 ft 8 in (4,47 m); wing area, 264·5 sq ft (24,57 m²).

GENERAL DYNAMICS F-16
FIGHTING FALCON

Country of Origin: USA.

Type: Single-seat air combat fighter (F-16A) and two-seat operational trainer (F-16B).

Power Plant: One 15,000 lb (6 805 kg) dry and 23,830 lb (10 809 kg) reheat Pratt & Whitney F100-PW-200 turbofan.

Performance: Max. speed (with wingtip AAMs), 1,350 mph (2 170 km/h) at 40,000 ft (12 190 m) or Mach 2·05, 915 mph (1 472 km/h) at sea level or Mach 1·2; tactical radius (interdiction mission HI-LO-HI on internal fuel with six Mk 82 bombs), 340 mls (550 km).

Weights: Operational empty, 14,567 lb (6 613 kg); max. take-off (air–air with wingtip AAMs), 23, 357 lb (10 594 kg), (interdiction), 35,400 lb (16 057 kg).

Armament: One 20-mm M61A-1 multi-barrel rotary cannon and external ordnance load of 12,000 lb (5 443 kg) distributed between nine stations (two wingtip, six underwing and one fuselage), or 15,200 lb (6 894 kg) with reduced internal fuel.

Status: First of two (YF-16) prototypes flown January 20, 1974, and first of eight pre-series aircraft on December 8, 1976. First full-production F-16A flown August 7, 1978, and 300 (F-16As and Bs) delivered by beginning of 1981 from Fort Worth (14 monthly), Schiphol, Netherlands (three monthly) and Gosselies, Belgium (two monthly). At beginning of 1981, USAF planning called for 1,388 (including 204 F-16Bs), with 124 for Netherlands, 116 for Belgium, 58 for Denmark, 72 for Norway, 75 for Israel and 40 for Egypt.

Notes: An export version, the F-16/J79 powered by an 18,730 lb (8 496 kg) General Electric J79-GE-119 turbojet, flew as a prototype on October 29, 1980 under a company-funded programme. The two-seat F-16B is illustrated above.

GENERAL DYNAMICS F-16 FIGHTING FALCON

Dimensions: Span (excluding missiles), 31 ft 0 in (9,45 m); length, 47 ft 7¾ in (14,52 m); height, 16 ft 5¼ in (5,01 m); wing area, 300 sq ft (27,87 m²).

GRUMMAN E-2C HAWKEYE

Country of Origin: USA.

Type: Shipboard airborne early warning, surface surveillance and strike control aircraft.

Power Plant: Two 4,910 ehp Allison T56-A-425 turboprops.

Performance: Max. speed (at max. take-off), 348 mph (560 km/h) at 10,000 ft (3 050 m); max. range cruise, 309 mph (498 km/h); max. endurance, 6·1 hrs; mission endurance (at 230 mls/370 km from base), 4·0 hrs; ferry range, 1,604 mls (2 580 km); initial climb, 2,515 ft/min (12,8 m/sec); service ceiling, 30,800 ft (9 390 m).

Weights: Empty, 38,009 lb (17 240 kg); max. take-off, 51,900 lb (23 540 kg).

Accommodation: Crew of five comprising flight crew of two and Airborne Tactical Data System team of three, each at an independent operating station.

Status: First of two E-2C prototypes flown on January 20, 1971, with first production aircraft flying on September 23, 1972. Sixty-four E-2Cs delivered by beginning of Fiscal Year 1981 against total US Navy requirement for 79 plus four for Israel and four ordered by Japan in September 1979 for delivery 1982—83, with a follow-on order for a further four anticipated.

Notes: Current production model of Hawkeye following 59 E-2As (all subsequently updated to E-2B standard) and, since December 1976, equipped with advanced APS-125 radar processing system. This system is able to operate independently, in co-operation with other aircraft, or in concert with ground environments. A training version currently serving with the US Navy is designated TE-2C, and production of the E-2C Hawkeye is scheduled to continue through 1986.

114

GRUMMAN E-2C HAWKEYE

Dimensions: Span, 80 ft 7 in (24,56 m); length, 57 ft 7 in (17,55 m); height, 18 ft 4 in (5,69 m); wing area, 700 sq ft (65,03 m²).

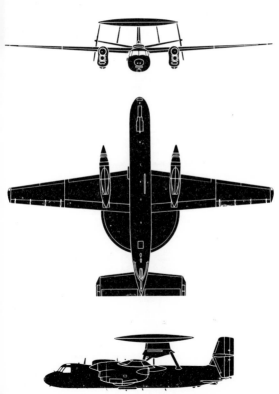

GRUMMAN F-14A TOMCAT

County of Origin: USA.

Type: Two-seat shipboard multi-purpose fighter.

Power Plant: Two 20,900 lb (9 480 kg) reheat Pratt & Whitney TF30-P-412A turbofans.

Performance: Design max. speed (clean), 1,545 mph (2 486 km/h) at 40,000 ft (12 190 m) or Mach 2·34; max. speed (internal fuel and four AIM-7 missiles at 55,000 lb/ 24 948 kg), 910 mph (1 470 km/h) at sea level or Mach 1·2; tactical radius (internal fuel and four AIM-7 missiles plus allowance for 2 min combat at 10,000 ft/3 050 m), approx. 450 mls (725 km); time to 60,000 ft (18 290 m) at 55,000 lb (24 948 kg), 2·1 min.

Weights: Empty equipped, 40,070 lb (18 176 kg); normal take-off (internal fuel and four AIM-7 AAMs), 55,000 lb (24 948 kg); max. take-off (ground attack/interdiction), 68,567 lb (31 101 kg).

Armament: One 20-mm M-61A1 rotary cannon and (intercept mission) six AIM-7E/F Sparrow and four AIM-9G/H Sidewinder AAMs or six AIM-54A and two AIM-9G/H AAMs.

Status: First of 12 research and development aircraft flown December 21, 1970, and some 385 had been delivered by the beginning of 1981 against anticipated US Navy requirement for 521, and 49 were in process of being fitted with a fuselage centreline tactical reconnaissance pod. An F-14 powered by two General Electric F101DFEs (considered as potential options to the TF30) is scheduled to fly July 1981.

GRUMMAN F-14A TOMCAT

Dimensions: Span (max.), 64 ft 1½ in (19,55 m), (min.), 37 ft 7 in (11,45 m), (overswept on deck), 33 ft 3½ in (10,15 m); length, 61 ft 11⅞ in (18,90 m); height, 16 ft 0 in (4,88 m); wing area, 565 sq ft (52,5 m²).

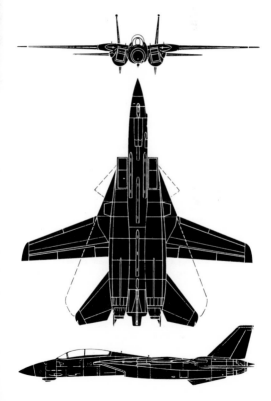

GULFSTREAM AMERICAN
GULFSTREAM III

Country of Origin: USA.
Type: Light business executive transport.
Power Plant: Two 11,400 lb (5 176 kg) Rolls-Royce Spey Mk 511-8 (RB.163-25) turbofans.
Performance: Max. cruising speed, 577 mph (928 km/h); long-range cruise, 512 mph (825 km/h); max. operational altitude, 45,000 ft (13 715 m); range (18 passengers and 45 min reserves), 4,375 mls (7 040 km) at long-range cruise, 3,080 mls (4 955 km) at max. cruise.
Weights: Operational empty, 38,300 lb (17 374 kg); max. take-off, 68,200 lb (30 936 kg).
Accommodation: Flight crew of two and various interior arrangements for 8–12 passengers as executive transport or up to 21 passengers with high-density interior.
Status: The Gulfstream III prototype was first flown on December 2, 1979, FAA certification was obtained on September 22, 1980, and firm orders for 70 had been placed by the beginning of 1981, when about a dozen had been delivered and production rate was two monthly, this being scheduled to increase to three monthly in 1983.
Notes: The Gulfstream III is a progressive development of the Gulfstream II (see 1969 edition) of which 258 were built. Some of these are now being fitted with the newer-technology Gulfstream III wing as Gulfstream IIBs, the first retrofit aircraft being scheduled for mid-1981 delivery.

GULFSTREAM AMERICAN GULFSTREAM III

Dimensions: Span, 77 ft 10 in (23,72 m); length, 83 ft 1 in (25,30 m); height, 24 ft 4½ in (7,40 m); wing area, 934·6 sq ft (86,82 m²).

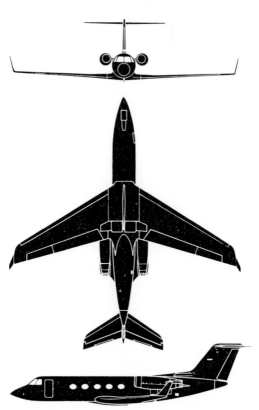

GULFSTREAM AMERICAN
PEREGRINE 600

Country of Origin: USA.
Type: Side-by-side two-seat basic trainer.
Power Plant: One 3,000 lb (1 360 kg) Pratt & Whitney JTD15D-5 turbofan.
Performance: (Estimated) Max. speed, 454 mph (730 km/h) at 20,000 ft (6 095 m); mission radius, 530–1,245 mls (850–2 000 km) at 40,000 ft (12 190 m); initial climb, 5,200 ft/min (26,4 m/sec).
Weights: Max. take-off, 6,200 lb (2 812 kg).
Status: Prototype scheduled to fly first quarter 1981.
Notes: Based on the design of the Hustler 500 business executive transport (see 1980 edition), production plans for which were discontinued in 1980, the Peregrine 600 prototype will have side-by-side seating and a single JTD15D turbofan. Proposed variants include a version with tandem seating and optional power plant comprising paired 850 lb (385 kg) Williams WR 44 turbofans fed via individual lateral flush intakes over the wing trailing edge.

GULFSTREAM AMERICAN PEREGRINE 600

Dimensions: Span, 34 ft 5½ in (10,50 m); length, 38 ft 4 in
(11,68 m); height, 13 ft 5 in (4,08 m); wing area, 192·76 sq ft
(17,90 m²).

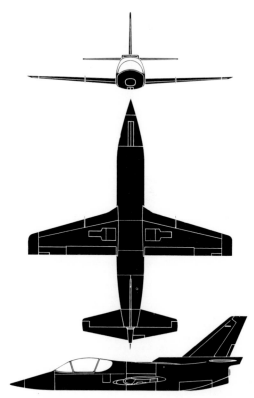

(HARBIN) Y-11

Country of Origin: Chinese Republic.

Type: Light utility transport and agricultural aircraft.

Power Plant: Two 285 hp Jia Hou-sai 6itsi (AI-14R) nine-cylinder radial air-cooled engines.

Performance: Max. speed, 137 mph (220 km/h); cruise (at 57% power), 102 mph (165 km/h); service ceiling, 13,120 ft (4 000 m); range, 618 mls (995 km); max. endurance, 7·5 hrs.

Weights: Empty, 4,520 lb (2 050 kg); max. take-off, 7,715 lb (3 500 kg).

Accommodation: Side-by-side seats on flight deck for pilot and co-pilot/passenger, and various arrangements for six—eight passengers in main cabin. Optional interiors providing for aero-medical role with four stretcher patients and one attendant.

Status: A prototype of the Y-11 was completed and flown at the State Aircraft Factory at Shenyang in 1975, a pre-series being initiated by the State Aircraft Factory at Harbin in 1978, and series production being launched by the latter in 1980, with some 25—30 having been delivered by beginning of 1981.

Notes: The Y-11, or Yun-shu 11 (Type 11 Transport Plane) was evolved primarily as a successor to the licence-manufactured Antonov An-2 (Y-5) biplane, and is a multi-role STOL aeroplane for rough-field operation, having both take-off and landing rolls of 153 yards (140 m). For the agricultural role (as illustrated) four insecticide atomisers are fitted under the wings, with two more attached to the stub wings, the fuselage accommodating a 1,885-lb (855-kg) powder hopper or a 214 Imp gal (973 l) tank. Two prototypes of a developed version, the Y-11T, are being built with 475 shp Pratt & Whitney PT6A-10 turboprops and featuring enlarged cabins for 16 passengers or 3,300 lb (1 500 kg) of freight.

(HARBIN) Y-11

Dimensions: Span, 55 ft 9¼ in (17,00 m); length, 39 ft 4½ in (12,00 m); height, 15 ft 2¾ in (4,64 m); wing area, 366 sq ft (34,00 m²).

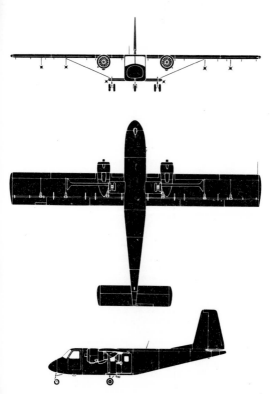

IAI KFIR-C2

Country of Origin: Israel.

Type: Single-seat multi-role fighter.

Power Plant: One 11,870 lb (5 385 kg) dry and 17,900 lb (8 120 kg) Bet-Shemesh-built General Electric J79-GE-17 turbojet.

Performance: (Estimated) Max. speed (50% fuel and two Shafrir AAMs), 850 mph (1 368 km/h) at 1,000 ft (305 m) or Mach 1·12, 1,420 mph (2 285 km/h) above 36,000 ft (10 970 m) or Mach 2·3; max. low-level climb rate, 47,250 ft/min (240 m/sec); max. ceiling, 59,050 ft (18 000 m); radius of action (air superiority mission with two 110 Imp gal/500 l drop tanks), 323 mls (520 km), (ground attack mission HI-LO-HI profile), 745 mls (1 200 km).

Weights: Loaded (intercept with 50% fuel and two AAMs), 20,700 lb (9 390 kg); max. take-off, 32,190 lb (14 600 kg).

Armament: Two 30-mm DEFA cannon with 125 rpg and (intercept) two or four Rafael Shafrir AAMs, or (ground attack) up to 8,820 lb (4 000 kg) of external ordnance.

Status: Initial production version of Kfir delivered to Israeli air arm from April 1975 with deliveries of improved Kfir-C2 having commenced early in 1977, production rate at the beginning of 1981 reportedly being 2·5 aircraft monthly.

Notes: The Kfir-C2 differs from the initial production Kfir (Young Lion) in having modifications designed primarily to improve combat manœuvrability, these comprising canard auxiliary surfaces which result in a close-coupled canard configuration, dog-tooth wing leading-edge extensions and nose strakes. Equipped with a dual-mode ranging radar, the Kfir is based on the Mirage 5 airframe, and a two-seat version was under development at the beginning of 1981.

IAI KFIR-C2

Dimensions: Span, 26 ft 11½ in (8,22 m); length, 51 ft 0¼ in (15,55 m); height, 13 ft 11½ in (4,25 m); wing area (excluding canard and dog-tooth), 375·12 sq ft (34,85 m²).

IAI WESTWIND 2

Country of Origin: Israel.
Type: Light business executive transport.
Power Plant: Two 3,700 lb (1 678 kg) Garrett AiResearch TFE 731-3-1G turbofans.
Performance: Max. speed, 533 mph (858 km/h) at 29,000 ft (8 840 m); econ. cruise, 449 mph (723 km/h) at 39,000–41,000 ft (11 890–12 500 m); range (four passengers), 3,345 mls (5 383 km), (10 passengers), 2,752 mls (4 429 km); max. initial climb, 5,000 ft/min (25,39 m/sec).
Weights: Empty equipped, 13,250 lb (6 010 kg); max. take-off, 23,500 lb (10 660 kg).
Accommodation: Two seats side-by-side on flight deck with various arrangements in main cabin for 7–10 passengers.
Status: The Westwind was first flown on April 24, 1979, and customer deliveries commenced in September 1980, production proceeding at the beginning of 1981 in parallel with the Westwind 1 at a combined rate of four monthly.
Notes: The Westwind 2 is a longer-range derivative of the Westwind 1 with a higher-efficiency aerofoil, winglets attached to the wingtip tanks and various other refinements. The Westwind 1, which succeeded the Westwind 1124 (the first of the Westwind series with TFE 731 turbofans), offered increases in fuel and cabin capacity, and some 130 TFE 731-powered Westwinds had been delivered by the beginning of 1981. At that time, work was proceeding on a further development, the Astra, which, for delivery in 1984, will retain the fuselage, power plant and tail unit of the Westwind 2, marrying these with an entirely new wing located beneath the cabin floor.

IAI WESTWIND 2

Dimensions: Span, 44 ft 9½ in (13,65 m); length, 52 ft 3 in (15,93 m); height, 15 ft 9½ in (4,81 m); wing area, 308·26 sq ft (28,64 m²).

ILYUSHIN IL-76 (CANDID)

Country of Origin: USSR.

Type: Heavy commercial and military freighter.

Power Plant: Four 26,455 lb (12 000 kg) Soloviev D-30KP turbofans.

Performance: Max. cruise, 497 mph (800 km/h) at 29,530 ft (9 000 m); range cruise, 466 mph (750 km/h) at 39,370 ft (12 000 m); max. range (with reserves), 4,163 mls (6 700 km); range with max. payload (88,185 lb/40 000 kg), 2,107 mls (5 000 km).

Weights: Max. take-off, 374,790 lb (170 000 kg).

Accommodation: Normal flight crew of four with navigator below flight deck in glazed nose. Pressurised hold for containerised and other freight. Military version has pressurised tail section for sighting 23-mm cannon barbette.

Status: First of four prototypes flown on March 25, 1971, with production deliveries to Soviet Air Force commencing 1974, and to Aeroflot (Il-76T) 1976. Export deliveries began in 1978 to Iraqi Airways (Il-76T), followed in 1979 by Libyan Arab Airlines (Il-76M) and Syrianair (Il-76M).

Notes: The Il-76 is being manufactured in both military and commercial versions, the former being illustrated opposite. Two commercial versions are current, the Il-76T, which introduced increased fuel capacity and higher take-off weights than the initial model, and the Il-76M (illustrated above) embodying unspecified modifications. A flight refuelling tanker version of the Il-76 is expected to achieve operational status with the Soviet Air Force in 1981–82, and an airborne warning and control system variant is known to be under development.

ILYUSHIN IL-76 (CANDID)

Dimensions: Span, 165 ft 8⅓ in (50,50 m); length, 152 ft 10¼ in (46,59 m); height, 48 ft 5⅛ in (14,76 m); wing area, 3,229·2 sq ft (300,00 m²).

ILYUSHIN IL-86 (CAMBER)

Country of Origin: USSR.

Type: Medium-haul commercial transport.

Power Plant: Four 28,660 lb (13 000 kg) Kuznetsov NK-86 turbofans.

Performance: Max. cruise, 590 mph (950 km/h) at 29,530 ft (9 000 m); long-range cruise, 559 mph (900 km/h) at 36,090 ft (11 000 m); range (with max. payload—350 passengers), 2,485 mls (4 000 km), (with 250 passengers), 3,107 mls (5 000 km).

Weights: Max. take-off, 454,150 lb (206 000 kg).

Accommodation: Standard flight crew of three–four and up to 350 passengers in basic nine-abreast seating with two aisles (divided between three cabins accommodating 111, 141 and 98 passengers respectively). Provision is made for passengers to load their own baggage into underfloor vestibules before entering the cabin via internal stairways. A version is proposed without this feature, which, with internal staircases deleted, will provide accommodation for a total of 375 passengers.

Status: First prototype flown on December 22, 1976, and production prototype flown on October 24, 1977. Service entry (with Aeroflot) is scheduled for early 1981, and production is a collaborative effort with Polish WSK-Mielec concern (complete stabiliser, all movable aerodynamic surfaces and engine pylons, and the entire wing were being built in Poland from 1980).

Notes: The first wide-body airliner of Soviet design, the Il-86 has been evolved under the supervision of General Designer G. V. Novozhilov and is intended for use on both domestic and international high-density routes. Four are to be supplied to LOT Polish Airlines in 1981.

130

ILYUSHIN IL-86 (CAMBER)

Dimensions: Span, 157 ft 8⅛ in (48,06 m); length, 195 ft 4 in (59,54 m); height, 51 ft 10½ in (15,81 m); wing area, 3,550 sq ft (329,80 m²).

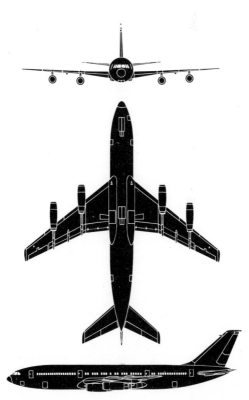

LEAR FAN MODEL 2100

Country of Origin: USA (United Kingdom*).
Type: Light business executive transport.
Power Plant: Two 825 shp (flat rated at 650 shp) Pratt & Whitney PT6B-35F turboshafts.
Performance: (Estimated) Max. cruising speed, 400 mph (644 km/h) at 30,000 ft (9 145 m); long-range cruise, 350 mph (563 km/h); range (at 6,000 lb/2 722 kg), 2,300 mls (3,700 km) at long-range cruise, (with max. payload), 2,070 mls (3 330 km); max. climb rate, 3,550 ft/min (18 m/sec); service ceiling, 41,000 ft (12 495 m).
Weights: Empty, 3,850 lb (1 746 kg); max. take-off, 7,200 lb (3 266 kg).
Accommodation: Pilot and co-pilot/passenger and six–seven passengers in main cabin, or eight in high-density configuration.
Status: First of three prototypes was flown on January 3, 1981. Series production to be undertaken by Lear Fan Ltd in Ulster with first series aircraft by late 1982, and deliveries commencing in following year. Orders totalled 175 aircraft by beginning of 1981.
Notes: The Model 2100 is radical both in concept and construction. The largest composite aircraft built to date, it uses graphite/epoxy (woven graphite impregnated with epoxy resin) construction for the fuselage and all surfaces, carries all fuel in integral wing tanks and has twin turboshafts driving a single pusher propeller via a gearbox drive train. All engineering, research and development is being undertaken at Reno, Nevada.

LEAR FAN MODEL 2100

Dimensions: Span, 39 ft 4 in (11,99 m); length, 39 ft 7 in (12,06 m); height, 11 ft 6 in (3,50 m); wing area, 162·9 sq ft (15,13 m²).

LOCKHEED C-130H-30 HERCULES

Country of Origin: USA.

Type: Medium- to long-range military transport.

Power Plant: Four 4,508 ehp Allison T56-A-15 turboprops.

Performance: Max. cruising speed (at 120,000 lb/ 54 430 kg), 361 mph (581 km/h) at 20,000 ft (6 100 m); range (with max. payload and 45 min reserves), 2,005 mls (3 226 km), (zero payload), 4,833 mls (7 778 km); max. initial climb, 1,900 ft/min (9,65 m/sec).

Weights: Operational empty, 73,181 lb (33 194 kg); max. normal take-off, 155,000 lb (70 310 kg); max. overload, 175,000 lb (79 380 kg).

Accommodation: Crew of four on flight deck comprising pilot, co-pilot, navigator and systems manager, and 128 troops, 92 fully-equipped paratroops, 93 casualty stretchers (plus six medical attendants), or seven cargo pallets.

Status: A "stretched" military version of the basic C-130H was offered by Lockheed–Georgia for the first time in 1980 as the C-130H-30 with initial deliveries of five to Indonesia during course of year. Hercules production (all versions) had just exceeded 1,600 at beginning of 1981, when production rate was three per month.

Notes: The C-130H-30 (like the commercial L-100-30) features two fuselage plugs totalling 180 in (4,57 m), and the UK Ministry of Defence is in process of converting 30 of the RAF's C-130K Hercules C Mk 1s (equivalent to C-130Hs) to "stretched" C-130H-30 standard as Hercules C Mk 3s. Variants of the "stretched" military Hercules currently offered by Lockheed–Georgia include the C-130H-30MP maritime surveillance and rescue version. Projected Hercules with further body stretch compromise the L-100-220 and -260 with 220-in (5,59-m) and 260-in (6,60-m) lengthening respectively.

134

LOCKHEED C-130H-30 HERCULES

Dimensions: Span, 132 ft 7 in (40,41 m); length, 112 ft 9 in (34,37 m); height, 38 ft 3 in (11,66 m); wing area, 1,745 sq ft (162,12 m²).

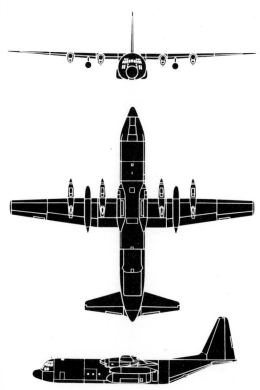

LOCKHEED P-3C ORION

Country of Origin: USA.

Type: Long-range maritime patrol aircraft.

Power Plant: Four 4,910 eshp Allison T56-A-14W turbo-props.

Performance: Max. speed at 105,000 lb (47 625 kg), 473 mph (761 km/h) at 15,000 ft (4 570 m); normal cruise, 397 mph (639 km/h) at 25,000 ft (7 620 m); patrol speed, 230 mph (370 km/h) at 1,500 ft (457 m); loiter endurance (all engines) at 1,500 ft (457 m), 12.3 hrs, (two engines), 17 hrs; max. mission radius, 2,530 mls (4 075 km), with 3 hrs on station at 1,500 ft (457 m), 1,933 mls (3 110 km); initial climb, 2,880 ft/min (14,6 m/sec).

Weights: Empty, 61,491 lb (27 890 kg); normal max. take-off, 133,500 lb (60 558 kg); max. overload, 142,000 lb (64 410 kg).

Accommodation: Normal flight crew of 10 of which five housed in tactical compartment. Up to 50 combat troops and 4,000 lb (1 814 kg) of equipment for trooping role.

Armament: Weapons bay can house two Mk 101 depth bombs and four Mk 43, 44 or 46 torpedoes, or eight Mk 54 bombs. External ordnance load of up to 13,713 lb (6 220 kg).

Status: YP-3C prototype flown October 8, 1968, P-3C deliveries commencing to US Navy mid-1969 with some 200 by 1981 against planned procurement (through 1989) of 316. Licence manufacture by Kawasaki of 42 (of 45) for Japanese Maritime Self-Defence Force, six to Iran (as P-3Fs), 10 to the RAAF, 18 for Canada (as CP-140 Auroras) with completion by March 1981, and 13 for the Netherlands with deliveries commencing from late 1981.

Notes: The Canadian version combines the P-3 airframes and engines with the electronic systems of the carrier-based S-3A Viking (see 1978 edition).

LOCKHEED P-3C ORION

Dimensions: Span, 99 ft 8 in (30,37 m); length, 116 ft 10 in (35,61 m); height, 33 ft 8½ in (10,29 m); wing area, 1,300 sq ft (120,77 m²).

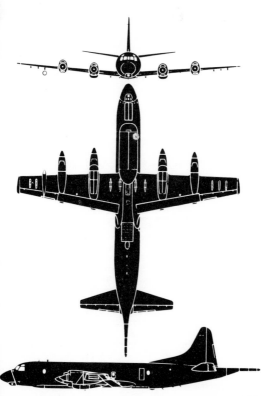

LOCKHEED L-1011-500 TRISTAR

Country of Origin: USA.

Type: Long-haul commercial transport.

Power Plant: Three 48,000 lb (21 772 kg) Rolls-Royce RB.211-524B turbofans.

Performance: (Estimated) Max. cruise, 608 mph (978 km/h) at 31,000 ft (9 450 m); econ. cruise, 567 mph (913 km/h) at 31,000 ft (9,450 m), or Mach 0·84; range (with full passenger payload), 6,053 mls (9 742 km), (with space limited max. payload), 4,855 mls (7 815 km).

Weights: Operational empty, 240,139 lb (108 925 kg); max. take-off, 496,000 lb (224 982 kg).

Accommodation: Basic flight crew of three and mixed-class arrangement for 222 economy (nine-abreast seating) and 24 first (six-abreast seating) class passengers.

Status: First L-1011-500 (for British Airways) flown on October 16, 1978. Total of 249 (all versions) on order (plus 56 options) at beginning of 1981, in which year 27 are scheduled for delivery (following 25 in 1980).

Notes: The TriStar 500 is a shorter-fuselage longer-range derivative of the basic L-1011-1 transcontinental version of the TriStar, a 62-in (157,5-cm) section being removed from the fuselage aft of the wing and a 100-in (254-cm) section forward. Versions with the standard fuselage are the L-1011-1, -100 and -200, the last-mentioned model (see 1977 edition) featuring additional centre section fuel tankage and -524 in place of -22B or -22F engines of 42,000 (19 050 kg) and 43,500 lb (19 730 kg) respectively. The -400 version is similar to the -500 but with the -1 wing and smaller engines.

LOCKHEED L-1011-500 TRISTAR

Dimensions: Span, 164 ft 3½ in (50,07 m); length, 164 ft 2 in (50,04 m); height, 55 ft 4 in (16,87 m); wing area, 3,541 sq ft (328,96 m²).

McDONNELL DOUGLAS DC-9 SUPER 80

Country of Origin: USA.

Type: Short- to medium-haul commercial transport.

Power Plant: Two 19,250 lb (8 730 kg) Pratt & Whitney JT8D-209 turbofans (alternative rating of 18,500 lb/8 400 kg).

Performance: Max. cruising speed, 577 mph (928 km/h) at 27,000 ft (8 230 m); long-range cruise, 508 mph (817 km/h) at 35,000 ft (10 670 m); range (with max. payload), 1,508 mls (2 427 km) at 523 mph (841 km/h) at 33,000 ft (10 060 m); max. range (with 22,760 lb/10 324 kg), 3,167 mls (5 095 km) at 508 mph (817 km/h) at 35,000 ft (10 670 m).

Weights: Operational empty, 78,666 lb (35 683 kg); max. take-off, 140,000 lb (63 503 kg).

Accommodation: Flight crew of two and typical mixed-class accommodation for 23 first-class and 137 economy-class passengers, or 155 all economy or 172 commuter-type layouts with five-abreast seating.

Status: The first Super 80 was flown on October 18, 1979, with first customer delivery (to Swissair) on September 12, 1980. By the beginning of 1981, a total of 87 (plus 23 options) Super 80s were on order for 12 customers, orders for DC-9s of all types totalling 1,071 at that time, the 1,000th DC-9 having been delivered in July 1980.

Notes: The Series 80 is the most recent and largest member of the DC-9 family, having a 14 ft 3 in (4,34 m) fuselage 'stretch' by comparison with what was previously the largest DC-9, the Series 50, coupled with an extended wing with new leading-edge slats, uprated engines and larger tailplane. The basic version is the Super 81 (described above), the Super 82 having 20,000 lb (9 072 kg) JT8D-217 turbofans and a 147,000-lb (66 680-kg) take-off weight, the Super 83 being a proposed European charter version with a 152,000-lb (68 945-kg) take-off weight, and the Super 80SF is a projected short-field version with Series 40 fuselage (see 1972 edition).

McDONNELL DOUGLAS DC-9 SUPER 80

Dimensions: Span, 107 ft 10 in (32,85 m); length, 147 ft 10 in (45,08 m); height, 29 ft 4 in (8,93 m); wing area, 1,279 sq ft (118,8 m²).

McDONNELL DOUGLAS DC-10 SERIES 30

Country of Origin: USA.

Type: Medium-range commercial transport.

Power Plant: Three 52,500 lb (23 814 kg) General Electric CF6-50C1 turbofans.

Performance: Max. cruise (at 400,000 lb/181 440 kg), 594 mph (956 km/h) at 31,000 ft (9 450 m); long-range cruise, 540 mph (870 km/h) at 31,000 ft (9 450 m); range (with max. payload), 6,195 mls (9 970 km) at 575 mph (925 km/h) at 31,000 ft (9 450 m); max. range, 7,400 mls (11 910 km) at 540 mph (870 km/h).

Weights: Operational empty, 261,459 lb (118 597 kg); max. take-off, 572,000 lb (259 457 kg).

Accommodation: Flight crew of three plus provision on flight deck for two supernumerary crew. Typical mixed-class accommodation for 225–270 passengers. Max. authorised passenger accommodation, 380 (plus crew of 11).

Status: First DC-10 (Series 10) flown August 29, 1970, with first Series 30 (46th DC-10 built) flying June 21, 1972, being preceded on February 28, 1972, by first Series 40. Orders (including KC-10A—see page 150) totalled 362 by 1981.

Notes: The DC-10 Series 30 and 40 have identical fuselages to the DC-10 Series 10 (see 1972 edition) and 15, but whereas these last-mentioned versions are domestic models, the Series 30 and 40 are intercontinental models and differ in power plant, weights and wing details, and in the use of three main under-carriage units. The Series 40 has 53,000 lb (24 040 kg) Pratt & Whitney JT9D-59A turbofans but is otherwise similar to the Series 30. The Series 10 and 15 have 41,000 lb (18 597 kg) CF6-6s and 46,500 lb (21 090 kg) CF6-45B2s respectively.

McDONNELL DOUGLAS DC-10 SERIES 30

Dimensions: Span, 165 ft 4 in (50,42 m); length, 181 ft 4¾ in (55,29 m); height, 58 ft 0 in (17,68 m); wing area, 3,921·4 sq ft (364,3 m²).

McDONNELL DOUGLAS AV-8B

Country of Origin: USA.
Type: Single-seat V/STOL ground attack aircraft.
Power Plant: One 21,500 lb (9 760 kg) Rolls-Royce F402-RR-405 vectored-thrust turbofan.
Performance: (Estimated FSD aircraft) Max. speed (clean aircraft), 685 mph (1 100 km/h) at 1,000 ft (305 m) or Mach 0·9, (with high-drag external load), 530 mph (852 km/h) at 5,000 ft (1 525 m) or Mach 0·71; combat radius (with max. external load and one gun), 213 mls (343 km), (with full internal fuel, seven 580-lb/259-kg bombs and one gun), 247 mls (398 km), (with two external tanks), 725 mls (1 167 km); ferry range (four drop tanks), 2,880 mls (4 635 km).
Weights: Operational empty, 12,750 lb (5 783 kg); max. short take-off, 28,750 lb (13 041 kg); max. take-off, 29,750 lb (13 495 kg).
Armament: Provision for two pod-mounted 30-mm cannon beneath fuselage. Six wing pylons and fuselage centreline pylon for up to 9,200 lb (4 173 kg), typical max. load comprising 16 570-lb (259-kg) Mk 82 bombs.
Status: Derived from Harrier (see pages 30–31) alias AV-8A, first of two YAV-8Bs flown November 9, 1978. Work proceeding at beginning of 1981 on four full-scale development (FSD) AV-8Bs with first scheduled to fly in October 1981 (illustrated opposite). US Marine Corps has requirement for 336 aircraft of this type.

McDONNELL DOUGLAS F-15C EAGLE

Dimensions: Span, 42 ft 9¾ in (13,05 m); length, 63 ft 9 in (19,43 m); height, 18 ft 5½ in (5,63 m); wing area, 608 sq ft (56,50 m²).

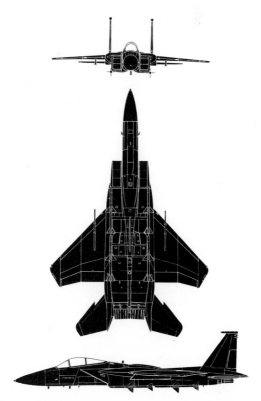

McDONNELL DOUGLAS F-18A HORNET

Country of Origin: USA.

Type: Single-seat shipboard fighter and attack aircraft.

Power Plant: Two 10,600 lb (4810 kg) dry and 16,000 lb (7 260 kg) reheat General Electric F404-GE-400 turbofans.

Performance: Max. speed (with two wingtip-mounted AIM-9s and two fuselage-mounted AIM-7s), 1,190 mph (1 915 km/h) above 36,000 ft (10 970 m) or Mach 1·8, 915 mph (1 472 km/h) at sea level or Mach 1·2; combat radius (fighter escort mission on internal fuel), 460 mls (740 km), (with three 315 US gal/1 192 l external tanks), 735 mls (1 180 km).

Weights: Loaded (air superiority mission with two AIM-9s, two AIM-7s and full cannon ammunition), 35,800 lb (16 240 kg); max. take-off, 50,000 lb (22 680 kg).

Armament: One 20-mm multi-barrel M61 rotary cannon with 570 rounds and (air–air combat) two IR-homing AIM-9 Sidewinder and two radar-guided AIM-7 Sparrow missiles. Nine external stores stations for maximum of 17,500 lb (7 940 kg) ordnance with full internal fuel.

Status: First of 11 (nine F-18As and two TF-18As) full-scale development (FSD) aircraft flown November 18, 1978, and remainder by second quarter of 1980. Six production aircraft delivered by beginning of 1981, when 105 had been ordered against planned procurement of 1,366 (including 153 TF-18As) for US Navy and USMC. Shore-based version to be procured by Canadian Armed Forces which will receive 113 single-seat CF-18As and 24 two-seat CF-18Bs commencing October 1982.

Notes: Hornet is to equip 24 US Navy attack and six fighter squadrons, and 20 USMC attack and 12 fighter squadrons.

McDONNELL DOUGLAS F-18A HORNET

Dimensions: Span, 37 ft 6 in (11,43 m); length, 56 ft 0 in (17,07 m); height, 15 ft 4 in (4,67 m); wing area, 400 sq ft (37,16 m²).

McDONNELL DOUGLAS KC-10A EXTENDER

Country of Origin: USA.

Type: Flight refuelling tanker and cargo aircraft.

Power Plant: Three 52,500 lb (23 814 kg) General Electric CF6-50C2 turbofans.

Performance: (Estimated) Max. speed, 620 mph (988 km/h) at 33,000 ft (10 060 m); max. cruise, 595 mph (957 km/h) at 31,000 ft (9 450 m); long-range cruise, 540 mph (870 km/h) at 31,000 ft (9 450 m); typical refuelling mission, 2,200 mls (3 540 km) from base with 200,000 lb (90 720 kg) of fuel and return; max. range (with 170,000 lb/77 112 kg cargo), 4,370 mls (7 033 km).

Weights: Operational empty (tanker), 239,747 lb (108 749 kg), (cargo configuration), 243,973 lb (110 660 kg); max. take-off, 590,000 lb (267 624 kg).

Accommodation: Flight crew of five plus provision for six seats for additional crew and four bunks for crew rest. Fourteen additional seats for support personnel may be provided in the forward cabin. Alternatively, a larger area can be provided for 55 more support personnel, with necessary facilities, to make total accommodation for 80, including the flight crew.

Status: The first KC-10A was flown on July 12, 1980, and six had been ordered for the USAF by the beginning of 1981 against a total requirement for 26 aircraft.

Notes: The Extender is a military tanker/cargo derivative of the commercial DC-10 Series 30 convertible freighter with refuelling boom, boom operator's station, hose and drogue, military avionics and body fuel cells in the lower cargo compartments.

McDONNELL DOUGLAS KC-10A EXTENDER

Dimensions: Span, 165 ft 4 in (50,42 m); length, 182 ft 0 in (55,47 m); height, 58 ft 1 in (17,70 m); wing area, 3,647 sq ft (338,8 m²).

MICROTURBO MICROJET 200

Country of Origin: France.

Type: Staggered side-by-side two-seat basic trainer.

Power Plant: Two 202 lb (92 kg) Microturbo TRS-18-046 turbojets.

Performance: (Estimated) Max. cruising speed, 288 mph (463 km/h); range (10% reserves), 621 mls (1 000 km) at 276 mph (444 km/h) at 16,405 ft (5 000 m); initial climb (at 2,063 lb/936 kg), 2,953 ft/min (15,0 m/sec), (at 2,315 lb/ 1 050 kg), 2,165 ft/min (11,0 m/sec); service ceiling, 29,530 ft (9 000 m).

Weights: (Pre-series aircraft) Empty equipped, 1,323 lb (600 kg); max. take-off (single-seat version), 2,063 lb (936 kg), (two-seat version), 2,315 lb (1 050 kg).

Status: The prototype Microjet 200 was first flown on June 24, 1980, and the construction of three pre-series aircraft was proceeding at the beginning of 1981.

Notes: With the Caproni Vizzola C 22J (see pages 58–59), the Microjet 200 is one of the first of a new generation of lightweight jet trainers. The prototype is of wooden construction, but the pre-series aircraft will have metal fuselages and wings of plastic composite construction. Both single- and two-seat versions are proposed, the latter having a staggered side-by-side arrangement, with the right-hand seat positioned 21·6 in (55 cm) aft of that on the left which is occupied by the pupil. This permits some small reduction in fuselage cross section (by comparison with a conventional side-by-side arrangement) and is claimed to simulate more closely the forward cockpit of a tandem seater to which the pupil will graduate.

152

MICROTURBO MICROJET 200

Dimensions: Span, 24 ft 10¾ in (7,58 m); length, 19 ft 10⅛ in (6,05 m); height, 7 ft 1¾ in (2,18 m); wing area, 65·88 sq ft (6,12 m²).

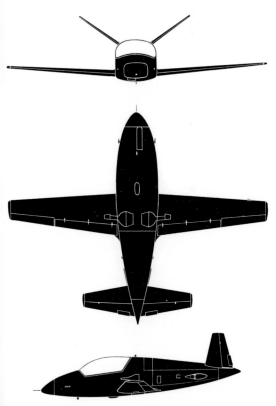

MIKOYAN MIG-23MF (FLOGGER-B)

Country of Origin: USSR.
Type: Single-seat (MiG-23MF) interceptor and (MiG-23BM)
strike fighter.
Power Plant: One 17,635 lb (8 000 kg) dry and 25,350 lb
(11 500 kg) reheat Tumansky R-29B turbofan.
Performance: Max. speed, 838 mph (1 350 km/h) at 1,000 ft
(305 m) or Mach 1·1, 1,520 mph (2 446 km/h) above
39,370 ft (12 000 m); combat radius (intercept mission with
four AAMs and 176 Imp gal/800 l centreline tank) 450–
500 mls (725–805 km); max range (with three 176 Imp gal/
800 l drop tanks), 1,400 mls (2 250 km) at 495 mph (795 km/h)
or Mach 0·75; ceiling, 60,000 ft (18 300 m).
Weights: Normal loaded (clean), 34,170 lb (15 500 kg); max.
take-off, 44,312 lb (20 100 kg).
Armament: One 23-mm twin-barrel GSh-23L cannon plus
(MiG-23MF) two AA-7 Apex and two AA-8 Aphid AAMs, or
(MiG-23BM) up to 9,920 lb (4 500 kg) of bombs and/or air-
to-surface missiles on five external stations.
Status: Prototype MiG-23 first flown early 1967, and initial
intercept model (MiG-23S) phased into SoVAF service from
1971. Current models include MiG-23MF, MiG-23BM and
BN, and two-seat MiG-23UM.
Notes: MiG-23BM differs from MiG-23MF Flogger-B (illustrated above and opposite) in having a redesigned nose with
the *High Lark* intercept radar replaced by radar ranging and
a laser ranger. Export versions of the MF and BN (with down-
graded equipment) are known as the Flogger-E and -F respectively, and production of the BN version is to be undertaken
in India 1982–83.

MIKOYAN MIG-23MF (FLOGGER-B)

Dimensions: (Estimated) Span (17 deg sweep), 46 ft 9 in (14,25 m), (72 deg sweep), 27 ft 6 in (8,38 m); length (including probe), 55 ft 1½ in (16,80 m); wing area, 293·4 sq ft (27,26 m²)

MIKOYAN MIG-25 (FOXBAT)

Country of Origin: USSR.

Type: Single-seat interceptor (Foxbat-A), high-altitude reconnaissance aircraft (Foxbat-B and -D) and two-seat conversion trainer (Foxbat-C).

Power Plant: Two 20,500 lb (9 300 kg) dry and 27,120 lb (12 300 kg) reheat Tumansky R-31 turbojets.

Performance: (Foxbat-A) Max. (dash) speed (with four AAMs), 1,850 mph (2 980 km/h) above 36,000 ft (10 970 m) or Mach 2·8; max. speed at sea level, 650 mph (1 045 km/h) or Mach 0·85; initial climb, 40,950 ft/min (208 m/sec); time to 36,000 ft (10 970 m), 2·5 min; service ceiling, 80,000 ft (24 000 m); mission radius (range-optimised profile), 590 mls (950 km), (dash only), 250 mls (400 km); max. ferry range, 1,600 mls (2 575 m).

Weights: Empty equipped, 44,100 lb (20 000 kg); max. take-off, 77,160 lb (35 000 kg).

Armament: Four AA-5 Ash or AA-6 Acrid AAMS (two infra-red homing and two semi-active radar homing).

Status: MiG-25 entered V-VS service (in Foxbat-A form) in 1970, recce versions following in 1971 (Foxbat-B) and 1973 (Foxbat-D). Production continuing at beginning of 1981.

Notes: The Foxbat-B and -D differ from the Foxbat-A (illustrated opposite) in having a new nose section, the -B (illustrated above) having a battery of live cameras and SLAR (side-looking radar) and the -D having the cameras omitted (being a dedicated electronic intelligence version) with a larger SLAR and other intelligence gathering equipment.

MIKOYAN MIG-25 (FOXBAT)

Dimensions: Span, 45 ft 9 in (13,94 m); length, 73 ft 2 in (22,30 m); height, 18 ft 4½ in (5,60 m); wing area, 602·8 sq ft (56,00 m²).

MIKOYAN MIG-27 (FLOGGER)

Country of Origin: USSR.

Type: Single-seat tactical strike fighter.

Power Plant: One 14,330 lb (6 500 kg) dry and 17,900 lb (8 125 kg) reheat Tumansky R-29-300 turbofan.

Performance: (Estimated) Max. speed, 724 mph (1 165 km/h) at 300 ft (90 m), or Mach 0·95, 1,056 mph (1 700 km/h) at 40,000 ft (12 200 m), or Mach 1·6; combat radius (LO-LO-LO mission profile with 176 Imp gal/800 l centreline tank and six 1,102-lb/500-kg bombs), 240 mls (390 km) at 610 mph (980 km/h), or Mach 0·8 (with dash and escape at Mach 0·95), (HI-LO-HI profile), 500 mls (805 km).

Weights: (Estimated) Normal loaded (clean), 35,000 lb (15 875 kg); max. take-off, 45,000 lb (20 410 kg).

Armament: One 23-mm six-barrel rotary cannon and six 1,102-lb (500-kg) bombs, or mix of AS-7 *Kerry*, AS-9, AS-11 and AS-12 ASMs, and rocket pods.

Status: Believed to have entered service with V-VS Frontal Aviation in 1975–76 (*Flogger-D*), with production continuing (*Flogger-J*) at beginning of 1981.

Notes: Dedicated tactical strike derivative of MiG-23 (see pages 154–155), sharing redesigned, drooped nose (to improve ground target acquisition) incorporating laser ranger with MiG-23BM (*Flogger-F*), but having simplified, fixed-ramp air intakes, additional armour, broader aft centre fuselage (to accommodate revised undercarriage for grassfield operation), shorter reheat nozzle, hardpoints on moving portion of wing and a Gatling-type cannon. Latest variant (*Flogger-J*) features further nose revision and equipment changes.

MIKOYAN MIG-27 (FLOGGER-D)

Dimensions: (Estimated) Span (17 deg sqeep), 46 ft 9 in (14,25 m), (72 deg sweep), 27 ft 6 in (8,38 m); length (including probe), 54 ft 0 in (16,46 m); wing area, 293·4 sq ft (27,26 m²).

MITSUBISHI MU-2B-60 MARQUISE

Country of Origin: Japan.

Type: Light business executive transport.

Power Plant: Two 715 shp Garrett AiResearch TPE 331-10-501M turboprops.

Performance: Max. cruising speed, 355 mph (571 km/h) at 16,000 ft (4 880 m); econ. cruise, 340 mph (547 km/h) at 20,000 ft (6 100 m); range (max. fuel), 1,606 mls (2 584 km) at 31,000 ft (9 450 m) with 45 min reserves; max. initial climb, 2,200 ft/min (18,29 mm/sec); service ceiling, 29,400 ft (8 960 m).

Weights: Empty equipped, 7,650 lb (3 470 kg); max. take-off, 11,575 lb (5 250 kg).

Accommodation: Seats for pilot and co-pilot/passenger on flight deck and various seating arrangements for seven—nine passengers in main cabin.

Status: The Marquise is a progressive refinement of the original MU-2 first flown on September 14, 1963, and since built in 15 versions, with sales (all models) approaching 700 by the beginning of 1981, when production was continuing of two versions in parallel, the Marquise, first flown on September 13, 1977, and the Solitaire (MU-2B-40), first flown on October 28, 1977.

Notes: The Solitaire differs from the Marquise (described here) in having 665 shp TPE 331 engines and a 6 ft 1 in (1,85 m) shorter fuselage with seating for six—seven in the main cabin. All airframe components are delivered to an assembly line at San Angelo, USA, where engines, avionics, furnishings, etc, are installed.

MITSUBISHI MU-2B-60 MARQUISE

Dimensions: Span, 39 ft 2 in (11,94 m); length, 39 ft 5 in (12,02 m); height 13 ft 8 in (4,17 m); wing area, 178 sq ft (16,55 m²).

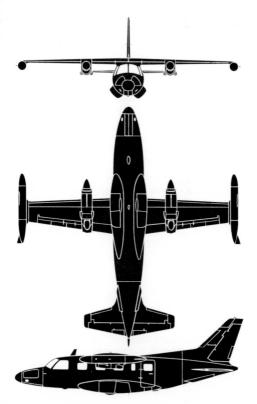

MITSUBISHI MU-300 DIAMOND I

Country of Origin: Japan.

Type: Light business executive transport.

Power Plant: Two 2,500 lb (1 134 kg) Pratt & Whitney JT15D-4 turbofans.

Performance: Max. operating speed, 500 mph (804 km/h) at 30,000 ft (9 145 m); typical cruise, 472 mph (760 km/h) at 39,000 ft (11 890 m); long-range cruise, 432 mph (695 km/h) at 39,000 ft (11 890 m); initial climb, 3,000 ft/min (15,2 m/sec); max. range (four passengers), 1,800 mls (2 896 km).

Weights: Empty equipped, 8,300 lb (3 764 kg); max. take-off, 13,890 lb (6 300 kg).

Accommodation: Pilot and co-pilot/passenger on flight deck and seven–nine passengers in main cabin.

Status: First of two prototypes flown on August 29, 1978, with second following in December 1978. The first four production aircraft were assembled and flown by Mitsubishi at Nagoya, sets of components for subsequent aircraft being shipped to the Mitsubishi facility at San Angelo, Texas, for assembly and flight testing. First customer deliveries were scheduled for March 1981, some 120 having been ordered by the beginning of the year. Production tempo is to attain five aircraft monthly by July 1981, six monthly by January 1982 and eight monthly by June 1982, with 130 delivered by the end of that year.

Notes: The Diamond is of thoroughly conventional structural design, the wing geometry being computer-generated and emphasis being placed on economy of operation, low noise level and good field performance, the last-mentioned attribute being obtained by means of large-span Fowler-type flaps, double-slotted on their inboard positions. Lateral control is by means of spoilers also used as speed brakes.

MITSUBISHI MU-300 DIAMOND I

Dimensions: Span, 43 ft 5 in (13,23 m); length, 48 ft 4 in
(14,73 m); height, 13 ft 9 in (4,19 m); wing area, 241·4 sq ft
(22,43 m²).

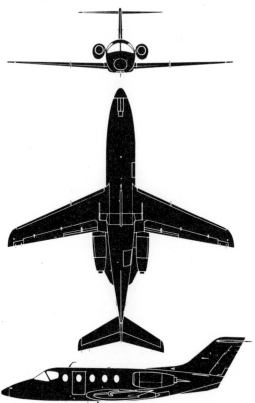

(NANCHENG) KIANG 5 (FANTAN-A)

Country of Origin: Chinese Republic.
Type: Single-seat tactical strike fighter.
Power Plant: Two 5,730 lb (2 600 kg) dry and 7,165 lb (3 250 kg) reheat Wopen 6A (modified Tumansky RD-9B-811) turbojets.
Performance: (Estimated) Max. speed, 900 mph (1 450 km/h) at 32,800 ft (10 000 m) or Mach 1·35, 685 mph (1 100 km/h) or Mach 0.95 at sea level; range cruise, 590 mph (950 km/h) or Mach 0·83; tactical radius (LO-LO-LO mission profile with two 167 Imp gal/760 I drop tanks and full internal ordnance), 230 mls (370 km), (HI-LO-HI), 405 mls (650 km); ferry range (max. external fuel), 1,270 mls (2 050 km).
Weights: (Estimated) Empty, 13,000 lb (5 900 kg); normal loaded, 17,000 lb (7 700 kg); max. take-off, 19,000 lb (8 620 kg).
Armament: Two 30-mm cannon in wing roots. Typical internal ordnance load, four 551-lb (250-kg) bombs, plus two 551-lb (250-kg) bombs on fuselage hardpoints and two or four pods each containing eight 57-mm rockets on wing stations.
Status: Prototype development in the late 'sixties with production deliveries commencing 1972–73 and continuing from Nancheng factory at beginning of 1981.
Notes: The Kiang 5 (A-5) interdictor and counter-air aircraft is a derivative of the Chinese-built MiG-19S (F-6) day fighter. It differs from its predecessor primarily in having a redesigned forward fuselage with lateral air intakes, a revised centre fuselage incorporating a weapons bay, a revised vertical tail and extended flaps. It was reported late 1980 that the Kiang 5 is now being offered for export.

(NANCHENG) KIANG 5 (FANTAN-A)

Dimensions: (Estimated) Span, 29 ft 6 in (9,00 m); length (without probe), 47 ft 0 in (14,30 m); height, 13 ft 0 in (3,95 m).

NORMAN NDN 6 FIELDMASTER

Country of Origin: United Kingdom.
Type: Two-seat agricultural monoplane.
Power Plant: One 750 shp Pratt & Whitney PT6A-34AG turboprop.
Performance: (Estimated) Max. speed, 188 mph (303 km/h); cruise (75% power), 171 mph (275 km/h); range (max. fuel and no reserves), 1,174 mls (1 889 km); max. initial climb, 1,200 ft/min (6,1 m/sec).
Weights: (Estimated) Empty equipped, 3,500 lb (1 588 kg); max. take-off, 8,500 lb (3 856 kg).
Status: First flight of prototype scheduled for third quarter of 1981, with production deliveries commencing early 1982.
Notes: Financed jointly by NDN Aircraft and the UK National Research Development Corporation, the Fieldmaster features an integral hopper which is part of the primary fuselage structure, carrying the engine bearers at its front end and the rear fuselage with cockpit aft. This hopper has a capacity of 581 Imp gal (2 642 l) and its integral construction permits a smaller cross section to be used for a given capacity. A liquid spray dispersal system is incorporated in a full-span auxiliary aerofoil flap, and all fuel is accommodated in the outer wing panels for maximum safety. A loader is accommodated on a "buddy" seat behind the pilot and removable dual controls will be available for flight training or check-out procedures. The Fieldmaster is claimed to be the first agricultural aircraft designed from the outset for the turboprop power.

NORMAN NDN 6 FIELDMASTER

Dimensions: Span, 50 ft 3 in (15,32 m); length, 36 ft 2 in (10,97 m); height, 11 ft 5 in (3,48 m); wing area, 338 sq ft (31,42 m²).

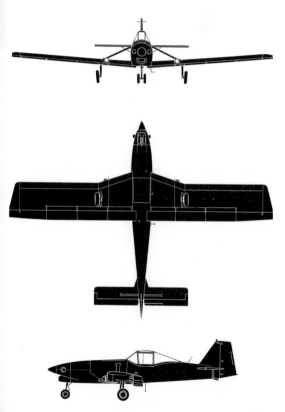

NORTHROP F-5E TIGER II

Country of Origin: USA.

Type: Single-seat air superiority, (RF-5E) tactical reconnaissance fighter and (F-5F) two-seat combat trainer.

Power Plant: Two 3,500 lb (1 588 kg) dry and 5,000 lb (2 268 kg) reheat General Electric J85-GE-21 turbojets.

Performance: Max. speed (at 13,220 lb/5 997 kg), 1,056 mph (1 700 km/h) or Mach 1·63 at 36,090 ft (11 000 m), 760 mph (1 223 km/h) or Mach 1·0 at sea level, (with wingtip missiles), 990 mph (1 594 km/h) or Mach 1·5 at 36,090 ft (11 000 m); combat radius (internal fuel), 173 mls (278 km), (with 229 Imp gal/1 041 l drop tank), 426 mls (686 km); initial climb (at 13,220 lb/5 997 kg), 31,600 ft/min (160,53 m/sec); combat ceiling, 53,500 ft (16 305 m).

Weights: Take-off (wingtip launching rail configuration), 15,400 lb (6 985 kg); max. take-off, 24,676 lb (11 193 kg).

Armament: (RF-5E) One or (F-5E) two 20-mm M-39 cannon and two wingtip-mounted AIM-9 Sidewinder AAMs. Up to 7,000 lb (3 175 kg) of ordnance may be carried on five stations for the attack role.

Status: First F-5E flown August 11, 1972 and first deliveries February 1973. First RF-5E flown January 29, 1979. Total of 1,080 Tiger II series aircraft (F-5E and two-seat F-5F) had been delivered by the beginning of 1981 when production was running at five per month.

Notes: The RF-5E is a tactical reconnaissance version of the F-5E dedicated fighter, with armament reduced by one cannon and a lengthened nose with 26 cu ft (0,75 m³) space for cameras and other sensors carried on quick-change platforms. The F-5G with a single General Electric F404 turbofan will commence flight test in 1982.

NORTHROP F-5E TIGER II

Dimensions: Span, 26 ft 8 in (8,13 m); length (including nose probe), 48 ft 2 in (14,68 m); height, 13 ft 4 in (4,06 m); wing area, 186 sq ft (17,30 m²).

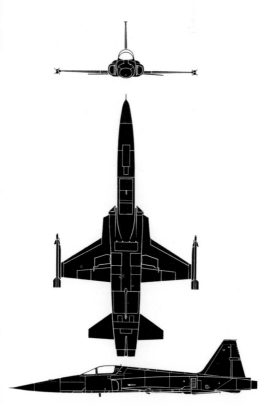

PANAVIA TORNADO F. Mᴋ 2

Country of Origin: United Kingdom.
Type: Tandem two-seat interceptor fighter.
Power Plant: Two 8,000 lb (3 623 kg) dry and 15,000 lb (6 810 kg) reheat Turbo-Union RB. 199-34R-04 Mk 101 (Improved) Turbofans.
Performance: (Estimated) Max. speed (clean) 840 mph (1 350 km/h) at 500 ft (150 m) or Mach 1·1, 1,450 mph (2 333 km/h) at 36,090 ft (11 000 m) or Mach 2·2; typical mission (with two external subsonic 330 Imp gal/1 500 l tanks), loiter for 2·0—2·5 hrs plus 10 min combat at 350—450 mls (560—725 km) from base; ferry range (full internal and max. external fuel), 2,000+ mls (3 200+ km).
Weights: Max. take-off (four Sky Flash and two Sidewinder AAMs plus two 330 Imp gal/1 500 l external tanks), 52,000 lb (23 587 kg).
Armament: One 27-mm Mauser cannon, two AIM-9L Sidewinder AAMs on inboard sides of swivelling wing pylons and four BAe Sky Flash AAMs in paired and staggered semi-recessed housings under fuselage.
Status: First of three prototype Tornado F. Mk 2s flown on October 27, 1979, with second and third following July 18 and November 18, 1980 respectively, the first production aircraft being scheduled for completion in 1983. RAF requirement for 165 aircraft with initial operational capability to be achieved by end of 1984.
Notes: The Tornado F. Mk 2 is a UK-only derivative of the multi-national (UK, Federal Germany and Italy) multi-role fighter (see 1978 edition) with Foxhunter intercept radar, more fuel in lengthened fuselage and a retractable air refuelling probe.

PANAVIA TORNADO F. Mk 2

Dimensions: Span (max.), 45 ft 7¼ in (13,90 m), (min.), 28 ft 2½ in (8,59 m); length, 59 ft 3 in (18,06 m); wing area, 322·9 sq ft (30,00 m²).

PILATUS PC-7 TURBO TRAINER

Country of Origin: Switzerland.

Type: Tandem two-seat basic trainer.

Power Plant: One 550 shp (flat rated from 650 shp) Pratt & Whitney PT6A-25A turboprop.

Performance: (At 4,189 lb/1 900 kg) Max. speed, 239 mph (385 km/h) at sea level, 264 mph (425 km/h) at 16,405 ft (5 000 m); cruise, 186 mph (300 km/h) at sea level, 205 mph (330 km/h) at 16,405 ft (5 000 m); max. range (at 40% power with 5% plus 20 min reserve), 777 mls (1 250 km); initial climb, 2,008 ft/min (10,2 m/sec).

Weights: Empty, 2,866 lb (1 300 kg); max. take-off (clean), 4,189 lb (1 900 kg), (external stores), 5,952 lb (2 700 kg).

Armament: Six wing hardpoints permit external loads up to maximum of 5,952 lb (2 700 kg).

Status: First of two PC-7 prototypes flown on April 12, 1966, and first production example flown July 1978. Initial production batch of 35 commenced in 1977, with follow-on batches laid down in 1978 and 1979. More than 160 on order for eight air arms (including Bolivia, Burma, Chile, Guatemala, Iraq and Mexico) by the beginning of 1981, when production rate was running at five per month, with approximately 80 aircraft delivered.

Notes: Derived from the piston-engined P-3 basic trainer, the prototypes being conversions of the original P-3 prototype and a series production P-3-05, the PC-7 has undergone extensive structural redesign in its production form and it is anticipated that this type will be adopted by the Swiss *Flugwaffe* which has a requirement for 40 aircraft for delivery 1981–83.

PILATUS PC-7 TURBO TRAINER

Dimensions: Span, 34 ft 1½ in (10,40 m); length, 31 ft 11⅞ in (9,75 m); height, 10 ft 6⅓ in (3,21 m); wing area, 178·68 sq ft (16,60 m²).

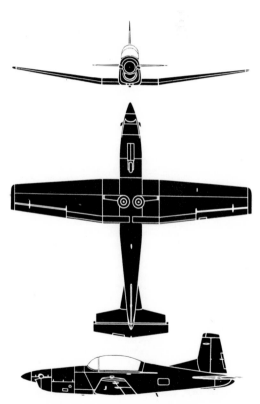

PILATUS BRITTEN-NORMAN
BN-2T TURBINE ISLANDER

Country of Origin: United Kingdom.

Type: Light utility transport.

Power Plant: Two 400 shp (derated to 320 shp) Allison 250-B17C turboprops.

Performance: Max. cruising speed, 180 mph (290 km/h) at sea level, 197 mph (317 km/h) at 10,000 ft (3 050 m); range (max. fuel and no reserves), 515 mls (829 km), (max. fuel with 10% reserve plus 45 min hold), 385 mls (620 km); initial climb, 1,100 ft/min (5,59 m/sec).

Weights: Empty equipped, 4,120 lb (1 869 kg); max. take-off, 6,000 lb (2 722 kg).

Accommodation: Flight crew of one or two and up to nine passengers (one beside pilot and four double seats).

Status: The BN-2T prototype was flown on August 2, 1980. Initial production aircraft scheduled for completion May 1981, with initial production rate of two per month.

Notes: The BN-2T is essentially a re-engined version of the BN-2B Islander (see 1980 edition), which, with 260 hp Avco Lycoming O-540-E4C5 (BN-2B-26/27) and 300 hp IO-540-K1B5 six-cylinder horizontally-opposed engines, continues in production in parallel. The extended wingtip auxiliary fuel tank extensions of the BN-2B-21 and -27 piston-engined Islanders are not available to the BN-2T, and the prototype (as illustrated) was completed in military Defender configuration with nose radar, the overall length quoted (opposite) being for the standard model. An earlier turbine-powered version of the Islander (with Lycoming LTP 101 turboprops) was first flown on April 6, 1977, but was discontinued after testing.

PILATUS BRITTEN-NORMAN BN-2T TURBINE ISLANDER

Dimensions: Span, 49 ft 0 in (14,94 m); length, 35 ft 7¾ in (10,86 m); height, 12 ft 4¾ in (3,77 m); wing area, 325 sq ft (30,19 m²).

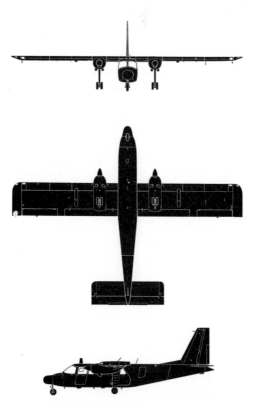

PIPER PA-31T CHEYENNE II XL

Country of Origin: USA.

Type: Light business executive transport.

Power Plant: Two 620 shp Pratt & Whitney PT6A-135 turbo-props.

Performance: Max. cruising speed, 317 mph (510 km/h) at 16,000 ft (4 875 m), 306 mph (492 km/h) at 21,000 ft (6 400 m); range (at max. range power), 1,590 mls (2 560 km) at 21,000 ft (6 400 m), (at max. cruise power), 1,170 mls (1 883 km) at 16,000 ft (4 875 m), 1,300 mls (2 090 km) at 21,000 ft (6 400 m); max. climb, 2,700 ft/min (13,7 m/sec).

Weights: Empty equipped, 5,680 lb (2 581 kg); max. take-off, 9,540 lb (4 336 kg).

Accommodation: Pilot and co-pilot/passenger on side-by-side individual seats and cabin seating for four—six passengers in individual seats.

Status: A stretched version of the Cheyenne II, the Cheyenne II XL began flight testing in 1979, the first production example being completed summer 1980, and customer deliveries being scheduled for April 1981.

Notes: The Cheyenne II XL differs from the Cheyenne II, which has been in production since 1973, in having a 24-in (61-cm) stretch in the forward cabin area and an additional window in the portside. The useful load is increased by more than 330 lb (150 kg). The 500th Cheyenne turboprop-powered light twin was delivered on August 13, 1980, this total including the less-powerful, lower-cost Cheyenne I, deliveries of which began in April 1978, and the first few examples of the larger Cheyenne III (see pages 182–183).

PIPER PA-31T CHEYENNE II XL

Dimensions: Span, 42 ft 8¼ in (13,01 m); length, 36 ft 8 in (11,10 m); height, 12 ft 9 in (3,89 m); wing area, 229 sq ft (21,30 m²).

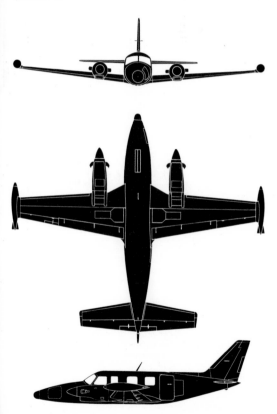

PIPER PA-32-301T TURBO SARATOGA

Country of Origin: USA.

Type: Light cabin monoplane.

Power Plant: One 300 hp Avco Lycoming TIO-540-S1AD six-cylinder horizontally-opposed turbo-supercharged engine.

Performance: Max. speed, 205 mph (329 km/h) at optimum altitude; cruise (75% power), 190 mph (306 km/h), (at 55% power), 152 mph (245 km/h); range (75% power), 787 mls (1 267 km), (55% power), 990 mls (1 594 km); initial climb, 1,075 ft/min (5,46 m/sec).

Weights: Standard empty, 2,000 lb (907 kg); max. take-off, 3,600 lb (1 633 kg).

Accommodation: Two individual seats for pilot and co-pilot/passenger with dual controls as standard and four reclining passenger seats with options for club arrangement (facing seats) and for a seventh seat between the centre pair. Two baggage compartments ahead of cockpit and aft of cabin.

Status: Introduced in 1980 as a successor to the PA-32 Cherokee Six and PA-32R Cherokee Lance, with production (all versions) running at some 30 monthly at the beginning of 1981, with approximately 390 delivered during 1980.

Notes: The Saratoga family comprises four members sharing a common airframe, these being the basic PA-32-301 and the turbo-supercharged -301T (described above), both of which feature a fixed undercarriage, and the PA-32R-301 and -301T which are similar but have fully-retractable undercarriages. These aircraft are available with a variety of factory-installed avionics packages.

PIPER PA-32-301T TURBO SARATOGA

Dimensions: Span, 36 ft 2 in (11,02 m); length, 28 ft 4 in (8,64 m); height, 8 ft 2½ in (2,50 m); wing area, 174·5 sq ft (16,20 m²).

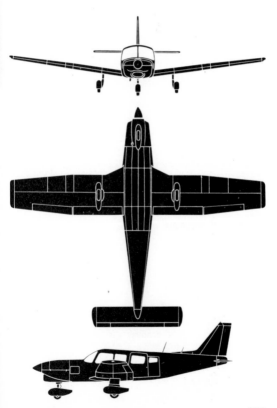

PIPER PA-38 TOMAHAWK

Country of Origin: USA.
Type: Side-by-side two-seat primary trainer.
Power Plant: One 112 bhp Avco Lycoming O-235-L2C four-cylinder horizontally-opposed engine.
Performance: Max. speed, 130 mph (209 km/h) at sea level; cruise (75% power), 125 mph (202 km/h) at 8,800 ft (2 680 m), (65% power), 117 mph (189 km/h) at 11,500 ft (3 505 m); range (with 45 min reserve), 463 mls (745 km) at 75% power, 500 mls (807 km) at 65% power; initial climb, 700 ft/min (3,55 m/sec); service ceiling, 12,850 ft (3 917 m).
Weights: Empty equipped, 1,064 lb (483 kg); max. take-off, 1,670 lb (757 kg).
Status: The PA-38 Tomahawk trainer was announced in October 1977, and customer deliveries commenced early in 1978, and more than 1,000 were delivered in the first year of production, some 220 being delivered during 1980, and production averaging 20 monthly at the beginning of 1981.
Notes: Placing emphasis on simplicity of maintenance and low operating costs, the Tomahawk incorporates a high degree of component interchangeability and several design features considered innovative in aircraft of its category. Like the Beechcraft Skipper (see pages 40–41), with which the Piper trainer is directly competitive, the Tomahawk employs a T-tail, which is claimed to afford greater stability and more positive rudder control, and its high aspect ratio wing of constant chord and thickness utilises a NASA Whitcomb aerofoil.

PIPER PA-38 TOMAHAWK

Dimensions: Span, 34 ft 0 in (10,36 m); length, 23 ft 1¼ in (7,04 m); height, 8 ft 7½ in (2,63 m); wing area, 125 sq ft (11,61 m²).

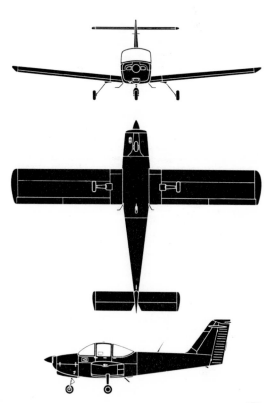

PIPER PA-42 CHEYENNE III

Country of Origin: USA.

Type: Light business executive transport.

Power Plant: Two 720 shp Pratt & Whitney PT6A-41 turboprops.

Performance: Max. speed, 336 mph (541 km/h) at 20,000 ft (6 095 m); max. continuous cruise, 333 mph (535 km/h) at 20,000 ft (6 095 m), 311 mph (500 km/h) at 33,000 ft (10 060 m); max. range (at max. range power), 2,348 mls (3 782 km) at 33,000 ft (10 060 m), (at max. cruise power), 2,055 mls (3 306 km) at 30,000 ft (9 150 m); initial climb, 2,400 ft/min (12,2 m/sec); service ceiling, 32,800 ft (10 000 m).

Weights: Empty, 6,389 llb (2 898 kg); max. take-off, 11,285 lb (5 129 kg).

Accommodation: Flight crew of one or two on separate flight deck and seating for six–nine in main cabin. Nose (300 lb/ 136 kg), aft cabin (300 lb/136 kg) and wing locker (300 lb/ 136 kg each) baggage compartments.

Status: Flight development of the Cheyenne III commenced in 1977, but this model subsequently underwent considerable further development, an extensively revised production proto-type flying on May 18, 1979, certification being achieved on December 18 of that year and customer deliveries commencing June 30, 1980.

Notes: The PA-42 is the largest member of the Cheyenne family of which the original PA-31 T prototype flew on August 20, 1969. The Cheyenne III should be compared with the Cheyenne II XL (pages 176–177).

PIPER PA-42 CHEYENNE III

Dimensions: Span, 47 ft 8⅛ in (14,53 m); length, 43 ft 4¾ in (13,23 m); height, 14 ft 9 in (4,50 m); wing area, 293 sq ft (27,20 m²).

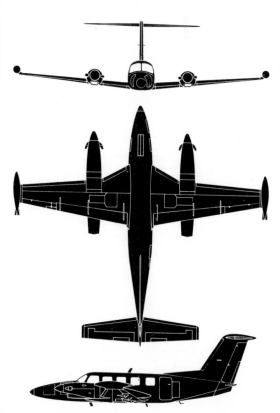

PZL M-18 DROMADER

Country of Origin: Poland.

Type: Single-seat agricultural aircraft.

Power Plant: One 987 hp PZL (Shvetsov) ASh-62IR nine-cylinder radial air-cooled engine.

Performance: (At 9,259 lb/4 200 kg) Max. speed (without agricultural equipment), 159 mph (256 km/h), (with spreader bar), 147 mph (237 km/h); cruise (without ag. equip.), 127 mph (205 km/h) at sea level, (with spreader bar), 118 mph (190 km/h); max. range (no reserves), 323 mls (520 km); initial climb (without ag. equip.), 1,142 ft/min (5,8 m/sec).

Weights: Empty equipped, 5,445 lb (2 470 kg); normal loaded, 9,529 lb (4 200 kg); max. take-off, 11,684 lb (5 300 kg).

Status: First and second prototypes flown on August 27 and October 2, 1976 respectively, and approximately 50 delivered by beginning of 1981 in both agricultural and fire-fighting versions.

Notes: The Dromader (Dromedary) has been developed by a PZL-Mielec team led by Jósef Oleksiak in collaboration with Rockwell International of the USA, and does, in fact, utilise Rockwell Thrust Commander outer wing panels. A fire-fighting version was first tested on November 11, 1978, and six of this variant have been exported to Canada. At the beginning of 1981, preparations were being made to install the Pratt & Whitney PT6A-45 turboprop in the Dromader. A 550 Imp gal (2 500 l) hopper is installed forward of the cockpit, and a Transland spreader for dusting with dry chemical or eight atomisers for fine spraying may be fitted.

PZL M-18 DROMADER

Dimensions: Span, 58 ft 0⅞ in (17,70 m); length, 31 ft 0⅞ in (9,47); height, 12 ft 1⅔ in (3,70 m); wing area, 430·56 sq ft (40,00 m²).

ROCKWELL COMMANDER JETPROP 1000

Country of Origin: USA.

Type: Light business executive transport.

Power Plant: Two 980 shp (flat rated to 733 shp) Garrett AiResearch TPE 331-10-501K turboprops.

Performance: Econ. cruising speed, 290 mph (467 km/h) at 31,000 ft (9 450 m); max. range (with 45 min reserves), 2,300 mls (3 700 km); initial climb, 2,804 ft/min (14,24 m/sec).

Weights: Empty equipped, 4,232 lb (1 920 kg); max. take-off, 11,200 lb (5 080 kg).

Accommodation: Two seats on flight deck for pilot and co-pilot/passenger and optional arrangements for six, eight or nine passenger seats in main cabin.

Status: The Commander Jetprop 1000 was introduced late in 1980 as a third member of the Jetprop Commander family of business executive aircraft, and customer deliveries were scheduled to commence early 1981, at which time production (all three models) was running at some 12 monthly, with some 95 delivered during 1980.

Notes: The name Commander Jetprop was adopted in 1980 for new variants of what was previously marketed as the Turbo Commander 690B, and there are three current production models all possessing the same overall dimensions. The Commander Jetprop 840 has engines (TPE 331-5-254Ks) flat rate at 717·5 shp from 840 shp and is otherwise similar to the Commander Jetprop 980. The latter has the same engines as the Commander Jetprop 1000 (described) which introduces a redesigned fuselage featuring more cabin space, increased accommodation and revised windows.

ROCKWELL COMMANDER JETPROP 1000

Dimensions: Span, 52 ft 1½ in (15,89 m); length, 42 ft 11½ in (13,10 m); height, 14 ft 11 in (4,56 m); wing area, 279·35 sq ft (25,95 m²).

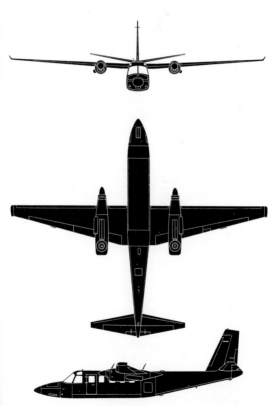

SAAB (JA) 37 VIGGEN

Country of Origin: Sweden.

Type: Single-seat all-weather intercepter fighter with secondary strike capability.

Power Plant: One 16,200 lb (7 350 kg) dry and 28,110 lb (12 750 kg) reheat Volvo Flygmotor RM 8B.

Performance: (Estimated) Max. speed (with two RB 24 Sidewinder AAMs), 1,320 mph (2 125 km/h) above 36,090 ft (11 000 m) or Mach 2·0, 910 mph (1 465 km/h) at 1,000 ft (305 m) or Mach 1·2; operational radius (M=2·0 intercept with two AAMs), 250 mls (400 km), (LO-LO-LO ground attack with six Mk 82 bombs), 300 mls (480 km); time (from brakes off) to 32,810 ft (10 000 m), 1·4 min.

Weights: (Estimated) Empty, 26,895 lb (12 200 kg); loaded (two AAMs), 37,040 lb (16 800 kg); max. take-off, 49,600 lb (22 500 kg).

Armament: One semi-externally mounted 30-mm Oerlikon KCA cannon with 150 rounds and up to 13,227 lb (6 000 kg) of ordnance on seven external stores stations.

Status: First of four JA 37 prototypes (modified from AJ 37 airframes) flown June 1974, with fifth prototype built from outset to JA 37 standards flown December 15, 1975. Initial production JA 37 flown on November 4, 1977. Total of 149 JA 37s (of 329 Viggens of all types) being procured.

Notes: The JA 37 is a development of the AJ 37 (see 1973 edition) which is optimised for the attack role. The JA 37 has uprated turbofan, carries a mix of BAe Sky Flash AAMs and cannon armament, and has X-band Pulse Doppler radar. The JA 37 attained initial operational capability mid-1980.

SAAB (JA) 37 VIGGEN

Dimensions: Span, 34 ft 9¼ in (10,60 m); length (excluding probe), 50 ft 8¼ in (15,45 m); height, 19 ft 4¼ in (5,90 m); wing area (including foreplanes), 561·88 sq ft (52,20 m²).

SEPECAT JAGUAR INTERNATIONAL

Countries of Origin: France and United Kingdom.
Type: Single-seat tactical strike fighter.
Power Plant: Two 5,320 lb (2 410 kg) dry and 8,040 lb (3 645 kg) reheat Rolls-Royce/Turboméca RT172-26 Adour 804, or 5,520 lb (2 504 kg) dry and 8,400 lb (3 811 kg) reheat RT172-58 Adour 811 turbofans.
Performance: (With -26 Adours and at typical weights) Max. speed, 820 mph (1 320 km/h) or Mach 1·1 at 1,000 ft (305 m), 1,057 mph (700 km/h) or Mach 1·6 at 32,810 ft (10 000 m); range (external fuel), 564 mls (907 km) LO-LO-LO, 875 mls (1 408 km) HI-LO-HI; ferry range, 2,190 mls (3 524 km).
Weights: Typical empty, 15,432 lb (7 000 kg); normal loaded, 24,000 lb (10 886 kg); max. take-off, 34,000 lb (15 422 kg).
Armament: Two 30-mm Aden or DEFA cannon and up to 10,000 lb (4 536 kg) ordnance on five external hardpoints. Provision for two AAMs on overwing pylons.
Status: The Jaguar International is an export version of the basic Jaguar, the first of eight prototypes of which was flown on September 8, 1968, 202 (including 37 two-seaters) having been delivered to the RAF and 185 (including 40 two-seaters) to the *Armée de l'Air*, with a further 15 remaining to be delivered to the latter at beginning of 1981. Twelve Jaguar Internationals delivered to each of Ecuador and Oman in 1977 and 1977–78 respectively, the latter having ordered a further 12 for delivery from early 1983, and delivery of 40 (including five two-seaters) to India commenced December 1980. Further 45 for India being supplied in knocked-down component form, with manufacture of further 60 by Hindustan Aeronautics.
Notes: Final 30 French aircraft equipped with Thomson-CSF Atlis 2 laser designator pod and AS 30L laser-guided AGM.

SEPECAT JAGUAR INTERNATIONAL

Dimensions: Span, 28 ft 6 in (8,69 m); length, 50 ft 11 in (15,52 m); height, 16 ft 0½ in (4,89 m); wing area, 260·3 sq ft (24,18 m²).

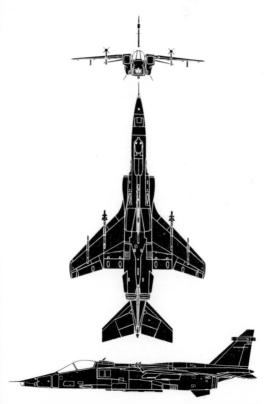

SHORTS 330

Country of Origin: United Kingdom.
Type: Third-level airliner and utility transport.
Power Plant: Two 1,173 shp Pratt & Whitney (Canada) PT6A-45A turboprops.
Performance: Max. cruise, 221 mph (356 km/h) at 10,000 ft (3 050 m); range cruise, 184 mph (296 km/h) at 10,000 ft (3 050 m); range (with 30 passengers and baggage, no reserve), 450 mls (725 km), (typical freighter configuration with 7,500-lb/3 400-kg payload), 368 mls (592 km); max. range (passenger configuration with 4,060-lb/1 840-kg payload), 1,013 mls (1 630 km), (freighter configuration with 5,400-lb/2 450-kg payload), 1,013 mls (1 630 km); max. climb, 1,210 ft/min (6,14 m/sec).
Weights: Empty equipped (for 30 passengers), 14,500 lb (6 577 kg); max. take-off, 22,400 lb (10 160 kg).
Accommodation: Standard flight crew of two and normal accommodation for 30 passengers in 10 rows three abreast and 1,000 lb (455 kg) of baggage.
Status: Engineering prototype flown August 22, 1974, with production prototype following on July 8, 1975. First production aircraft flown on December 15, 1975. Customer deliveries commenced mid-1976, and at the beginning of 1981 a total of 74 aircraft had been ordered with 60 delivered, and production rate was 1·5–2·0 monthly.
Notes: The Shorts 330 is derived from the Skyvan STOL utility transport (see 1975 edition) and is designed primarily for commuter and regional air service operators. Retaining many of the Skyvan's characteristics, including its large cabin cross section, Shorts 330 had been ordered by 23 operators in nine countries by the beginning of 1981.

SHORTS 330

Dimensions: Span, 74 ft 8 in (22,76 m); length, 58 ft 0½ in (17,69 m); height, 16 ft 3 in (4,95 m); wing area, 453 sq ft (42,10 m²).

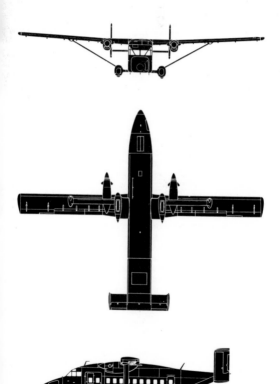

SHORTS 360

Country of Origin: United Kingdom.
Type: Third-level airliner and utility transport.
Power Plant: Two 1,294 shp Pratt & Whitney PT6A-65R turboprops.
Performance: (Estimated) Max. cruising speed, 243 mph (391 km/h) at 10,000 ft (3 050 m); range (with max. passenger payload), 265 mls (426 km) at max. cruise (with allowances for 100-mile/160-km diversion and 45 min hold), (with max. fuel), 655 mls (1 054 km) with same allowances.
Weights: Operational empty, 16,490 lb (7 480 kg); max. take-off, 25,700 lb (11 657 kg).
Accommodation: Crew of two on flight deck, plus cabin attendant, and standard seating for 36 passengers in 11 rows of two-plus-one with wide aisle and one row of three at rear of cabin. Baggage compartments in nose and aft of cabin.
Status: The prototype is scheduled to commence flight test during the last quarter of 1981, with certification and first commercial deliveries approximately one year later, the first announced customers being Suburban Airlines (for four) and Chautauqua Airlines (for two).
Notes: The Shorts 360 is a growth version of the Shorts 330 (see pages 192–193) differing from its progenitor primarily in having some 10% more power, a 3-ft (91-cm) rear "stretch" ahead of the wing and entirely redesigned rear fuselage and tail surfaces. The fuselage lengthening permits passenger capacity to be increased by two rows of three seats, and the uprated power plants and lower aerodynamic drag by comparison with the earlier aircraft result in higher cruising speeds.

SHORTS 360

Dimensions: Span, 74 ft 8 in (22,75 m); length, 70 ft 6 in (21,49 m); height, 22 ft 7 in (6,88 m); wing area, 453 sq ft (42,08 m²).

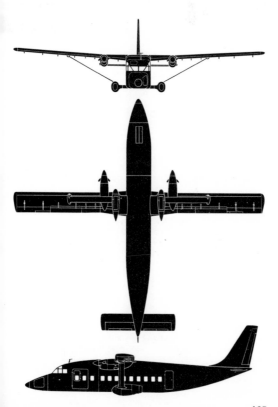

SIAI MARCHETTI S.211

Country of Origin: Italy.

Type: Tandem two-seat basic trainer.

Power Plant: One 2,500 lb (1 134 kg) Pratt & Whitney JT15D-4M turbofan.

Performance: (Estimated) Max. speed, 444 mph (715 km/h) at 25,000 ft (7 620 m); max cruise, 437 mph (703 km/h) at 25,000 ft (7 620 m); range (max. fuel and 30 min reserves), 1,186 mls (1 908 km) at 30,000 ft (9 145 m); max. initial climb, 4,950 ft/min (25,1 m/sec); service ceiling, 42,000 ft (12 800 m).

Weights: Empty equipped, 3,185 lb (1 445 kg); max. take-off (trainer), 5,070 lb (2 300 kg), (attack), 6,173 lb (2 800 kg).

Armament: (For armament training and attack) Four wing hardpoints stressed for loads up to 660 lbs (300 kg) inboard and 330 lb (150 kg) outboard, with max. external load of 1,320 lb (600 kg). Typical loads may include four 7,62-mm Minigun pods, four 12,7-mm gun pods, or (inboard only) two 20-mm cannon pods. Other options include four 18×50-mm, 6×68-mm or 7×2·75-in (70-mm) rocket pods, or four bombs of up to 330 lb (150 kg).

Status: First of two prototypes was scheduled to fly in March 1981, with second following in April—May.

Notes: The S.211 is being developed as a private venture and is intended to reverse the trend towards heavier, more complex and more expensive instructional aircraft. Siai Marchetti can offer production deliveries of the S.211 from late 1982. The S.211 is intended for basic and intermediate training up to the stage of more advanced trainers, such as the Alpha Jet and Hawk.

SIAI MARCHETTI S.211

Dimensions: Span, 26 ft 3 in (8,00 m); length, 30 ft 5½ in (9,28 m); height, 12 ft 2¾ in (3,73 m); wing area, 135·63 sq ft (12,60 m²).

SUKHOI SU-17 & SU-20 (FITTER)

Country of Origin: USSR.

Type: Single-seat attack aircraft and two-seat operational training and countermeasures aircraft.

Power Plant: One 17,195 lb (7 800 kg) dry and 24,700 lb (11 200 kg) reheat Lyulka AL-21F-3 turbojet.

Performance: Max. speed (clean), 808 mph (1 300 km/h) or Mach 1·06 at sea level, 1,430 mph (2 300 km/h) at 39,370 ft (12 000 m) or Mach 2·17; combat radius (LO-LO-LO mission profile), 260 mls (420 km), (HI-LO-HI), 373 mls (600 km); range (with 2,205-lb/1 000-kg ordnance and auxiliary fuel), 1,415 mls (2 280 km).

Weights: Max. take-off, 39,022 lb (17 700 kg).

Armament: Two 30-mm NR-30 cannon and (short-range mission) max. external ordnance load of 7,716 lb (3 500 kg). Loads include radio command guidance AS-7 Kerry, AS-9, AS-11 and AS-12 anti-radiation and AS-10 electro-optical ASMs.

Status: Variable-geometry derivative of fixed-wing Su-7 (Fitter-A), flown as prototype in 1966 as S-22I (alias Su-7IG), entered Soviet service in 1972 as Su-17 (Fitter-C) and progressively developed in both single- (Fitter-D and -H) and tandem two-seat (Fitter-E and -G) versions as dedicated interdiction and counter-air aircraft (Su-20). Production rate of 17–18 monthly at beginning of 1981.

Notes: Illustrated above and opposite in its Fitter-D version, this aircraft now has terrain-following radar and laser marked target seeker, and (as the Su-20 Fitter-F) has been supplied to Algeria, Egypt, Iraq, Libya and Poland. A less sophisticated version (the Su-22) has been supplied to Peru, South Yemen and Syria. The Fitter-G and -H have revised nose and dorsal contours, taller vertical surfaces and ventral fin.

198

SUKHOI SU-17 & SU-20 (FITTER)

Dimensions: (Estimated) Span (max.), 45 ft 0 in (13,70 m), (min.), 32 ft 6 in (9,90 m); length (including probe), 58 ft 3 in (17,75 m); height, 15 ft 5 in (4,70 m); wing area (max.), 410 sq ft (38,00 m²).

SUKHOI SU-24 (FENCER-A)

Country of Origin: USSR.

Type: Two-seat attack aircraft.

Power Plant: Two 17,635 lb (8 000 kg) dry and 25,350 lb (11 500 kg) Tumansky R-29 turbofans.

Performance: (Estimated) Max. speed (with two combat tanks), 915 mph (1 470 km/h) or Mach 1·2 at sea level, 1,520 mph (2 445 km/h) or Mach 2·3 above 36,000 ft (11 000 m); combat radius HI-LO-HI (with 4,400 lb/2 000 kg ordnance), 1,050 mls (1 690 km) with 593 mph (955 km/h) or Mach 0·9 cruise at 36,000 ft (11 000 m) with 115 mls (185 km) at 915 mph (1 473 km/h) or Mach 1·2 at sea level, LO-LO-LO, 345 mls (555 km) with 645 mph (1 035 km/h) or Mach 0·85 sea level cruise and same allowance for supersonic dash and escape; time to 40,000 ft (12 190 m) from brakes-off, 1·5 min.

Weights: (Estimated) Empty equipped, 41,890 lb (19 000 kg); max. take-off, 87,080 lb (39 500 kg).

Armament: One 23-mm six-barrel Gatling-type rotary cannon and (short-range interdiction mission), 22 220-lb (100-kg) or 551-lb (250-kg) bombs, or 16 1,102-lb (500-kg) bombs distributed between nine stores stations (including two swivelling pylons on movable wing panels). Alternative loads include radio command guidance AS-7 *Kerry* ASMs, AS-9, AS-11 and AS-12 anti-radiation ASMs and AS-10 electro-optical ASMs, plus AA-2-2 *Atoll* and AA-8 *Aphid* AAMs.

Status: Prototype of the Su-24 is believed to have flown in 1970, with initial service from late 1974. In excess of 500 in Soviet service by beginning of 1981, when production rate was eight–nine monthly.

Notes: The Su-24 was developed specifically for the interdiction and counter-air missions.

SUKHOI SU-24 (FENCER-A)

Dimensions: (Estimated) Span (max.), 56 ft 6 in (17,25 m), (min.), 33 ft 9 in (10,30 m); length, 65 ft 6 in (20,00 m); height, 18 ft 0 in (5,50 m); wing area, 452 sq ft (42,00 m²).

TRANSALL C.160

Countries of Origin: France and Federal Germany.

Type: Medium-range tactical transport.

Power Plant: Two 6,100 ehp Rolls-Royce Tyne RTy 20 Mk 22 turboprops.

Performance: (At 112,435 lb/51 000 kg) Max. speed, 322 mph (518 km/h) at 16,000 ft (4 875 m); econ. cruise, 282 mph (454 km/h) at 20,000 ft (6 100 m); range (with 35,274-lb/ 16 000-kg payload), 783 mls (1 260 km), (with 17,640-lb/ 8 000-kg payload), 2,734 mls (4 400 km), (with 6,614-lb/ 3 000-kg payload), 4,350 mls (7 000 km); ferry range, 5,500 mls (8 850 km).

Weights: Basic operations, 63,814 lb (28 946 kg); max. take-off, 108,245 lb (49 100 kg).

Accommodation: Crew of three and 62–88 paratroops, max. of 93 fully-equipped troops, up to 63 casualty stretchers and four medical attendants, or loads up to 37,500 lb (17 000 kg).

Status: First prototype flown February 25, 1963, and two further prototypes and 179 production aircraft built by time series manufacture terminated in 1972. Production resumed in 1978 against requirement for further 25 aircraft for *Armée de l'Air* plus three for Indonesian government, with first scheduled to fly September 1981, with production being one per month thereafter. Programme shared between France (Aérospatiale) and Germany (MBB and VFW).

Notes: The new series C.160 embodies a number of modifications. The flight crew complement has been reduced from four to three, the wing centre section now contains 1,980 Imp gal (9 000 l) of fuel, flight refuelling equipment is provided and the aircraft can also operate as a flight refuelling tanker, a new autopilot system has been introduced and the forward fuselage has been redesigned to eliminate the port freight door.

TRANSALL C.160

Dimensions: Span, 131 ft 3 in (40,00 m); length, 106 ft 4 in (32,40 m); height 38 ft 3 in (11,65 m); wing area, 1,723·3 sq ft (160,10 m²).

TUPOLEV TU-22M BACKFIRE-B

Country of Origin: USSR.

Type: Long-range strike and maritime recce-strike aircraft.

Power Plant: Two (estimated) 33,070 lb (15 000 kg) dry and 46,300 lb (21 000 kg) reheat Kuznetsov turbofans.

Performance: (Estimated) Max. (short-period) speed, 1,320 mph (2 125 km/h) at 39,370 ft (12 000 m) or Mach 2·0; max. sustained speed, 1,190 mph (1 915 km/h) at 39,370 ft (12 000 m) or Mach 1·8, 685 mph (1 100 km/h) at sea level or Mach 0·9; cruise, 495 mph (795 km/h) at sea level or Mach 0·65, 530 mph (850 km/h) at 39,370 ft (12 000 m) or Mach 0·8; unrefuelled combat radius (including 400 mls/250 km at Mach 1·8), 2,485 mls (4 000 km) for HI-LO-HI (200 mls/ 320 km at low altitude) profile.

Weights: (Estimated) Operational empty, 114,640 lb (52 000 kg); max. take-off, 260,000 lb (118 000 kg).

Armament: One (semi-recessed on fuselage centreline) or two (side-by-side on pylons beneath air intake trunks) AS-4 Kitchen or AS-6 Kingfish inertially-guided stand-off missiles, or internal bomb load of approx. 15,000 lb (6 800 kg). Remote-controlled twin 23-mm cannon tail barbette.

Status: Reported in prototype form in 1969, the Backfire apparently began to enter service with both the long-range element of the Soviet Air Force and the Soviet naval air arm in 1974, with combined total of 175–190 in service at the beginning of 1981, when production rate was reportedly rising from 2·5 to 3·5 monthly.

Notes: There has been some confusion as to the correct designation of this type, Soviet sources designating it Tu-22M and western intelligence sources referring to it as Tu-26.

TUPOLEV TU-22M BACKFIRE-B

Dimensions: (Estimated) Span (max.), 115 ft 0 in (35,00 m), (min.), 92 ft 0 in (28,00 m); length, 138 ft 0 in (42,00 m); height, 29 ft 6 in (9,00 m); wing area, 1,830 sq ft (170 m²).

TUPOLEV TU-154B-2 (CARELESS)

Country of Origin: USSR.

Type: Medium- to long-haul commercial transport.

Power Plant: Three 23,150 lb (10 500 kg) Kuznetsov NK-8-2U turbofans.

Performance: Max. cruise, 590 mph (950 km/h) at 31,000 ft (9 450 m); econ. cruise, 559 mph (900 km/h) at 36,090 ft (11 000 m); range (with max. payload—39,683 lb/18 000 kg), 1,710 mls (2 750 km), (with 160 passengers), 2,020 mls (3 250 km), (with 120 passengers), 2,485 mls (4 000 km).

Weights: Max. take-off, 211,644 lb (96 000 kg).

Accommodation: Crew of three on flight deck and basic arrangements for 160 single-class passengers in six-abreast seating, eight first-class and 150 tourist-class passengers, or (high-density) 169 passengers.

Status: Prototype Tu-154 flown on October 4, 1968, current production model being the Tu-154B (introduced by Aeroflot in 1976) of which approximately four per month were being manufactured at the beginning of 1981. More than 350 Tu-154s (all versions) are currently in service with Aeroflot and the Tu-154B has also been supplied to Malev (three) of Hungary.

Notes: The Tu-154B-2 combines the improvements introduced by the Tu-154A, the major changes in controls and systems, and slight increases in weights featured by the basic Tu-154B and Thomson-CSF/SFIM automatic flight control and navigation equipment. The wing spoilers have been extended in span and are now used for low-speed lateral control and passenger capacity has been increased by extending the usable cabin area rearwards, and an extra emergency exit has been added in each side of the fuselage. Various longer-range versions of the basic Tu-154 are known to be under study.

TUPOLEV TU-154B-2 (CARELESS)

Dimensions: Span, 123 ft 2½ in (37,55 m); length, 157 ft 1¾ in (47,90 m); height, 37 ft 4¾ in (11,40 m); wing area, 2,168·92 sq ft (201,45 m²).

VALMET L-70 MILTRAINER

Country of Origin: Finland.
Type: Side-by-side two-seat primary trainer.
Power Plant: One 200 hp Avco Lycoming AEIO-360-A1B6 four-cylinder horizontally-opposed engine.
Performance: Max. speed (at 2,204 lb/1 000 kg), 149 mph (240 km/h) at sea level, (at max. take-off weight), 143 mph (230 km/h); cruise (75% power), 130 mph (210 km/h) at sea level; initial climb, 1,122 ft/min (5,7 m/sec); service ceiling, 15,090 ft (4 600 m); range (max. payload and no reserves), 534 mls (860 km).
Weights: Empty equipped, 1,691 lb (767 kg); max. take-off (aerobatic), 2,293 lb (1 040 kg), (normal category), 2,756 lb (1 250 kg).
Status: L-70X prototype flown July 1, 1975, and first production aircraft flown December 1979, with initial deliveries against order for 30 for Finnish Air Force commencing October 13, 1980, with five delivered by beginning of 1981.
Notes: Assigned the name Vinka (Blast) by the Finnish Air Force, the Miltrainer is replacing the Saab Safir as the service's primary-basic trainer. Four hardpoints in the wings can lift external loads up to 660 lb (300 kg), suiting the Miltrainer for armament training or, when flown as a single-seater, the tactical support mission. The Miltrainer may also be flown as a four-seater.

VALMET L-70 MILTRAINER

Dimensions: Span, 32 ft 3¾ in (9,85 m); length, 24 ft 7¼ in (7,50 m); height, 10 ft 0⅓ in (3,31 m); wing area, 150·69 sq ft (14,00 m²).

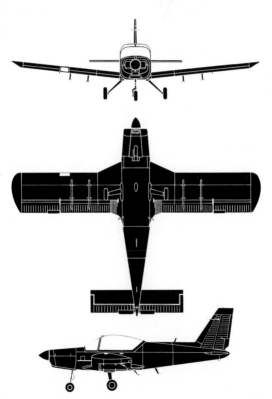

VOUGHT A-7K CORSAIR II

Country of Origin: USA.

Type: Tandem two-seat tactical strike fighter and operational proficiency aircraft.

Power Plant: One 15,000 lb (6 804 kg) Allison TF41-A-2 (Rolls-Royce Spey) turbofan.

Performance: Max. speed, 690 mph (1 110 km/h) at sea level, 685 mph (1 102 km/h) at 5,000 ft (1 525 m), (with 12 Mk 82 bombs), 646 mph (1 040 km/h) at 5,000 ft (1 525 m); combat range (with eight 800-lb/363-kg M117 bombs and two 250 Imp gal/1 136 l drop tanks), 1,340 mls (2 156 km) at average cruise of 508 mph (817 km/h); initial climb (at 42,000 lb/19 050 kg), 10,900 ft/min (55,3 m/sec).

Weights: Basic empty, 20,800 lb (9 435 kg); max. take-off, 42,000 lb (19 050 kg).

Armament: One 20-mm M61A-1 Vulcan rotary cannon in port side of fuselage and up to 14,000 lb (6 350 kg) of ordnance on eight (two fuselage and six wing) external stations.

Status: Prototype A-7K (converted from A-7D airframe) flown November 7, 1980. Twenty-four production A-7Ks ordered for Air National Guard by beginning of 1981 against total ANG requirement for 42 aircraft.

Notes: The A-7K is essentially a two-seat derivative of the single-seat A-7D with a similar 34-in (86,4-cm) fuselage plug (to accommodate the second cockpit) and marginally larger vertical tail surfaces to those of the US Navy's TA-7C (converted from A-7B and A-7C airframes with a Pratt & Whitney TF30-P-408 turbofan) and the Hellenic Air Force's TA-7H (similarly powered to the A-7K), the latter being a new-build two-seat version of the A-7H, which, in turn, is a Greek version of the A-7D.

VOUGHT A-7K CORSAIR II

Dimensions: Span, 38 ft 8¾ in (11,80 m); length, 48 ft 8 in (14,80 m); height, 16 ft 3⅜ in (4,97 m); wing area, 375 sq ft (34,83 m²).

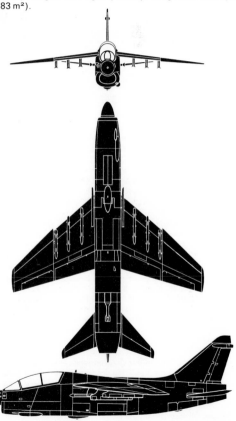

YAKOVLEV YAK-36MP (FORGER-A)

Country of Origin: USSR.

Type: Single-seat shipboard air defence and strike fighter.

Power Plant: One (approx.) 17,640 lb (8 000 kg) lift/cruise turbojet plus two 7,935 lb (3 600 kg) lift turbojets.

Performance: (Estimated) Max. speed, 695 mph (1 120 km/h) above 36,000 ft (10 970 m), or Mach 1·05, 725 mph (1 167 km/h) at sea level, or Mach 0·95; high-speed cruise, 595 mph (958 km/h) at 20,000 ft (6 095 m), or Mach 0.85; combat radius (internal fuel and 2,205-lb/1 000-kg external ordnance), 230 mls (370 km), (with two 110 Imp gal/500 l drop tanks, a reconnaissance pod and two AAMs), 340 mls (547 km); initial climb, 20,000 ft/min (101,6 m/sec).

Weights: (Estimated) Empty, 12,125 lb (5 500 kg); max. take-off, 22,000 lb (9 980 kg).

Armament: Four underwing pylons with total capacity of (approx.) 2,205 lb (1 000 kg), including twin-barrel 23-mm cannon pods, air-to-air missiles or bombs.

Status: The Yak-36MP (Forger-A) is believed to have flown in prototype form in 1971 and to have attained service evaluation status in 1976 aboard the carriers *Kiev* and *Minsk*.

Notes: Possessing no short-landing-and-take-off (STOL) capability, being limited to vertical-take-off-and-landing (VTOL operation), the Yak-36 combines a vectored-thrust lift/cruise engine with fore and aft lift engines. The single-seat Yak-36MP possesses no attack radar and no internal armament. A tandem two-seat version, the Yak-36UV (Forger-B), has an extended forward fuselage. A second seat is added ahead of that of the Yak-36MP and the nose is drooped to provide a measure of vertical stagger and the aft fuselage is extended.

YAKOVLEV YAK-36MP (FORGER-A)

Dimensions: (Estimated) Span, 24 ft 7 in (7,50 m); length, 52 ft 6 in (16,00 m); height, 11 ft 0 in (3,35 m); wing area, 167 sq ft (15,50 m²).

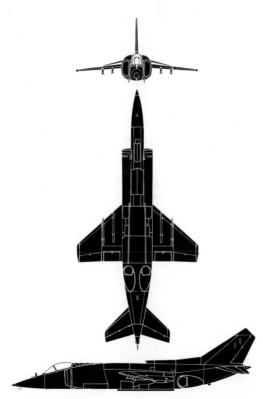

YAKOVLEV YAK-42 (CLOBBER)

Country of Origin: USSR.

Type: Short- to medium-haul commercial transport.

Power Plant: Three 14,320 lb (6 500 kg) Lotarev D-36 turbofans.

Performance: Econ. cruise, 510 mph (820 km/h) at 25,000 ft (7 600 m); range (max. payload—31,938 lb/14 500 kg), 620 mls (1 000 km), (with 26,430-lb/12 000-kg payload), 1,150 mls (1 850 km); max range, 1,520 mls (2 450 km); time to cruise altitude (25,000 ft/7 600 m), 11 min.

Weights: Operational empty, 63,845 lb (28 960 kg); max. take-off, 114,640 lb (52 000 kg).

Accommodation: Basic flight crew of two and various alternative cabin arrangements, including 76 passengers in a mixed-class layout (16 first class), 100 passengers in a single-class layout with six-abreast seating and 120 passengers in a high-density layout.

Status: First prototype flown on March 7, 1975, followed by second in April 1976. A production prototype was flown in February 1977, and deliveries to Aeroflot were originally scheduled to commence 1979, but were delayed until 1980, the first scheduled service having been anticipated late that year.

Notes: The initial prototypes of the Yak-42 differed one from the other in wing sweep angle, the first prototype featuring 11 deg of sweepback and the second 25 deg, the latter sweep angle being adopted for production aircraft. The Yak-42 is intended for operation primarily over relatively short stages and utilising restricted airfields with poor surfaces and limited facilities in the remoter areas of the Soviet Union.

YAKOVLEV YAK-42 (CLOBBER)

Dimensions: Span, 112 ft 2½ in (34,20 m); length, 119 ft 4 in (36,38 m); height, 32 ft 3 in (9,83 m); wing area, 1,615 sq ft (150,00 m²).

AÉROSPATIALE SA 330J PUMA

Country of Origin: France.

Type: Medium transport helicopter.

Power Plant: Two 1,575 shp Turboméca IVC turboshafts.

Performance: Max. speed, 163 mph (262 km/h); max. continuous cruise at sea level, 159 mph (257 km/h); max. inclined climb, 1,400 ft/min (7,1 m/sec); hovering ceiling (in ground effect), 7,315 ft (2 230 m), (out of ground effect), 4,430 ft (1 350 m); max. range (standard fuel), 342 mls (550 km).

Weights: Empty, 7,969 lb (3 615 kg); max. take-off, 16,534 lb (7 500 kg).

Dimensions: Rotor diam, 49 ft 5¾ in (15,08 m); fuselage length, 46 ft 1½ in (14,06 m).

Notes: The civil SA 330J and the equivalent military SA 330L (illustrated) were the current production models of the Puma at the beginning of 1981 when nearly 700 Pumas of all versions had been ordered. The SA 330J and 330L differ from the civil SA 330F (passenger) and SA 330G (cargo), and SA 330H (military) models that immediately preceded them in having new plastic blades accompanied by increases in gross weights. The SA 330B (French Army), SA 330C (export) and SA 330E (RAF) had 1,328 shp Turmo IIIC4 turboshafts. Components for the Puma are supplied by Westland in the UK (representing approx. 15% of the airframe) and production was five Pumas monthly at the beginning of 1981. The Puma has been delivered to some 45 countries.

AÉROSPATIALE AS 332L SUPER PUMA

Country of Origin: France.
Type: Medium transport helicopter (22 seats).
Power Plant: Two 1,780 shp Turboméca Makila turboshafts.
Performance: Max. cruising speed, 178 mph (286 km/h); econ. cruise, 158 mph (255 km/h); max. inclined climb, 1,890 ft/min (9,6 m/sec); hovering ceiling (in ground effect), 10,170 ft (3 100 m), (out of ground effect), 8,038 ft (2 450 m); range (standard fuel), 407 mls (655 km).
Weights: Empty, 8,708 lb (3 950 kg); max. take-off, 17,637 lb (8 000 kg).
Dimensions: Rotor diam, 49 ft 5¾ in (15,08 m); fuselage length, 48 ft 7¾ in (14,82 m).
Notes: The AS 332L, a stretched (by 2·5 ft/76,5 cm) version of the basic AS 332 Super Puma (see 1980 edition), was first flown on October 10, 1980. The AS 332L is the fourth of five Super Pumas that were engaged in the flight development programme at the beginning of 1981, the first SA 332 having flown on September 13, 1978. Four versions of the Super Puma are currently under development, the AS 332B and AS 332C being a military version accommodating 20 troops and a commercial version seating 17 passengers respectively, and the AS 332L and AS 332M which are respectively civil and military versions embodying the fuselage stretch and each providing four more seats.

217

AÉROSPATIALE SA 342 GAZELLE

Country of Origin: France.
Type: Five-seat light utility helicopter.
Power Plant: One 870 shp Turboméca Astazou XIVM turboshaft.
Performance: Max. speed, 193 mph (310 km/h); max. continuous cruise at sea level, 168 mph (270 km/h); max. inclined climb, 2,066 ft/min (10,5 m/sec); hovering ceiling (in ground effect), 13,120 ft (4 000 m), (out of ground effect), 10,330 ft (3 150 m); range at sea level, 488 mls (785 km).
Weights: Empty equipped, 2,149 lb (975 kg); max. take-off, 4,190 lb (1 900 kg).
Dimensions: Rotor diam, 34 ft 5½ in (10,50 m); fuselage length, 31 ft 2¾ in (9,53 m).
Notes: The SA 342 is a more powerful derivative of the SA 341 (592 shp Astazou IIIA) and has been exported to Kuwait, Iraq and elsewhere, and is equipped to launch four HOT missiles, AS-11s or other weapons. A civil equivalent, the SA 342J offering a 220 lb (100 kg) increase in payload, became available in 1977, and sales of the SA 341 and 342 Gazelles totalled some 900 by the beginning of 1981. Versions of the lower-powered SA 341 comprise the SA 341B (British Army), SA 341C (British Navy), SA 341D (RAF), SA 341F (French Army), SA 341G (civil version) and SA 341H (military export version). The latest military version is the SA 342M (illustrated) with a six-HOT installation, deliveries of which to the French Army began in 1980, 160 having been ordered.

AÉROSPATIALE AS 350 ECUREUIL

Country of Origin: France.

Type: Six-seat light general-purpose utility helicopter.

Power Plant: One (AS 350B) 641 shp Turboméca Arriel or (AS 350D) 615 shp Avco Lycoming LTS 101–600A2 turboshaft.

Performance: Max. speed, 169 mph (272 km/h) at sea level; cruise, 144 mph (232 km/h) at sea level; max. climb, 1,810 ft/min (9,2 m/sec); hovering ceiling (in ground effect), 9,678 ft (2 950 m), (out of ground effect), 7,382 ft (2 250 m); range, 435 mls (700 km) at sea level.

Weights: Empty, 2,304 lb (1 045 kg); max. take-off, 4,299 lb (1 950 kg).

Dimensions: Rotor diam, 35 ft 0¾ in (10,69 m); fuselage length, 35 ft 9½ in (10,91 m).

Notes: First (LTS 101-powered) Ecureuil (Squirrel) prototype flown on June 27, 1974, with second (Arriel-powered) following February 14, 1975. The Ecureuil is being manufactured with both types of power plant, the LTS 101-powered AS 350D being marketed in the USA as the AStar, some 500 having been ordered by North American customers by the beginning of 1981, when total orders for both versions exceeded 600 and production rate had reached 23 monthly. The standard model is a six-seater and features include a Starflex all-plastic rotor head, simplified dynamic machinery and modular assemblies to simplify changes in the field. More than 200 AS 350s had been delivered by the beginning of 1981.

AÉROSPATIALE AS 355E ECUREUIL 2

Country of Origin: France.
Type: Six-seat light general-purpose utility helicopter.
Power Plant: Two 425 shp Allison 250-C20F turboshafts.
Performance: Max. speed, 169 mph (272 km/h) at sea level; cruise, 149 mph (240 km/h) at sea level; max. climb, 1,710 ft/min (8,7 m/sec); hovering ceiling (in ground effect), 7,210 ft (2 200 m), (out of ground effect), 4,920 ft (1 500 m); range, 497 mls (800 km) at sea level.
Weights: Empty, 2,711 lb (1 230 kg); max. take-off, 4,630 lb (2 100 kg).
Dimensions: Rotor diam, 35 ft 0¾ in (10,69 m); fuselage length, 35 ft 9½ in (10,91 m).
Notes: Flown for the first time on September 27, 1979, the AS 355E Ecureuil 2 is a twin-engined derivative of the AS 350 (see page 219) intended primarily for North American customers to which it is marketed as the TwinStar. Claimed to be the cheapest and most compact twin-turboshaft helicopter available, more than 270 helicopters of this type had been ordered by customers in 10 countries by the beginning of 1981, when production schedules called for the commencement of deliveries early 1981, with 130 to be completed during year. The AS 355E Ecureuil 2 employs an essentially similar airframe and similar dynamic components to those of the AS 350 Ecureuil 1, including the composite material Starflex rotor head. Military versions are currently under development.

AÉROSPATIALE SA 361H DAUPHIN

Country of Origin: France.
Type: Light anti-armour helicopter.
Power Plant: One 1,400 shp Turboméca Astazou XXB turboshaft.
Performance: Max. cruising speed, 180 mph (289 km/h) at sea level; econ. cruise, 168 mph (270 km/h); max. climb, 2,885 ft/min (14,5 m/sec); hovering ceiling (in ground effect), 12,630 ft (3 850 m); range, 350 mls (565 km).
Weights: Max. take-off, 7,496 lb (3 400 kg).
Dimensions: Rotor diam, 38 ft 4 in (11,68 m); fuselage length, 40 ft 3 in (12,27 m).
Notes: The SA 361H/HCL (*hélicoptère de combat léger*) is an anti-armour version of the basic SA 361H military variant of the Dauphin (its civil equivalent being the SA 361F). It is equipped with a forward-looking infra-red aiming system and up to eight HOT (High-subsonic Optically-guided Tube-launched) anti-armour missiles, but retains its capability to transport up to 13 fully-equipped troops. The first prototype Dauphin flew on June 2, 1972, production being initiated as the SA 360, and the prototype of the SA 361, an overpowered version intended specifically for hot-and-high operating conditions, followed on July 12, 1976. If ordered, the HCL model can be delivered in 1982. The SA 360 Dauphin is powered by a 1,050 shp Astazou XVIIIA and normally carries a pilot and nine passengers.

AÉROSPATIALE SA 365N DAUPHIN 2

Country of Origin: France.

Type: Multi-purpose and transport helicopter.

Power Plant: Two 701 shp Turboméca Arriel 1 C turboshafts.

Performance: Max. speed, 196 mph (315 km/h) at sea level; cruise, 169 mph (272 km/h) at sea level; max. climb, 1,692 ft/min (8,6 m/sec); hovering ceiling (in ground effect), 6,070 ft (1 850 m), (out of ground effect), 3,455 ft (1 050 m); range, 525 mls (845 km) at sea level.

Weights: Empty, 4,162 lb (1 888 kg); max. take-off, 7,935 lb (3 600 kg).

Dimensions: Rotor diam, 38 ft 3½ in (11,68 m); fuselage length, 37 ft 4 in (11,38 m).

Notes: The SA 365N, the prototype of which flew on March 31, 1979, is the latest variant of the Dauphin 2 and is intended to supplant the SA 365C (see 1979 edition) in production with deliveries commencing in the first half of 1981. The SA 366G, first flown July 23, 1980, powered by Avco Lycoming LTS 101-750 turboshafts is a version selected by the US Coast Guard which plans procurement of 90 from 1982 onwards, and a combined production rate (SA 365N and SA 366G) of 10–15 monthly is anticipated for 1982, with 19 SA 365Ns delivered by the end of 1981. Accommodating a pilot and up to 13 passengers, the SA 365N differs from the SA 365C in having uprated Arriel engines, a reprofiled fuselage, a fully-retractable undercarriage and increased fuel capacity.

AGUSTA A 109A

Country of Origin: Italy.
Type: Eight-seat light utility helicopter.
Power Plant: Two 420 shp Allison 250-C20B turboshafts.
Performance: (At 5,402 lb/2 450 kg) Max. speed, 192 mph (310 km/h); max. continuous cruise, 173 mph (278 km/h) at sea level; hovering ceiling (in ground effect), 9,800 ft (2 987 m), (out of ground effect), 6,700 ft (2 042 m); max. inclined climb, 1,600 ft/min (8,12 m/sec); max. range, 385 mls (620 km) at 148 mph (238 km/h).
Weights: Empty equipped, 2,998 lb (1 360 kg); max. take-off, 5,780 lb (2 622 kg).
Dimensions: Rotor diam, 36 ft 1 in (11,00 m); fuselage length, 35 ft 2½ in (10,73 m).
Notes: The first of four A 109A prototypes flew on August 4, 1971. A pre-production batch of 10 A 109As was followed by first customer deliveries late 1976 with 115 delivered by beginning of 1981, when production was running at four–five machines monthly. The A 109A is currently being offered for civil and military roles, five having been delivered to the Italian Army, including two equipped to launch TOW (Tube-launched Optically-tracked Wire-guided) missiles. Variants include a naval A 109A with search radar, gyro-stabilised weapons aiming sight and torpedo or rocket armament, and an electronic warfare version with active electronic countermeasures and passive electronic suppression equipment.

BELL MODEL 206B JETRANGER III

Country of Origin: USA.

Type: Five-seat light utility helicopter.

Power Plant: One 420 shp Allison 250-C20B turboshaft.

Performance: (At 3,200 lb/1 451 kg) Max. speed, 140 mph (225 km/h) at sea level; max. cruise, 133 mph (214 km/h) at sea level; hovering ceiling (in ground effect), 12,700 ft (3 871 m), (out of ground effect), 6,000 ft (1 829 m); max. range (no reserve), 360 mls (579 km).

Weights: Empty, 1,500 lb (680 kg); max. take-off, 3,200 lb (1 451 kg).

Dimensions: Rotor diam., 33 ft 4 in (10,16 m); fuselage length, 31 ft 2 in (9,50 m).

Notes: Introduced in 1977, with deliveries commencing in July of that year, the JetRanger III differs from the JetRanger II which it supplants in having an uprated engine, an enlarged and improved tail rotor mast and more minor changes. Some 2,500 commercial JetRangers had been delivered by the beginning of 1981, both commercial and military versions (including production by licensees) totalling more than 6,300. A light observation version of the JetRanger for the US Army is designated OH-58 Kiowa and a training version for the US Navy is known as the TH-57A SeaRanger. The JetRanger is built by Agusta in Italy as the AB 206, and at the beginning of 1981, Agusta was producing the JetRanger at a rate of six monthly and the parent company was producing 20 monthly.

BELL MODEL 206L-1 LONGRANGER II

Country of Origin: USA.
Type: Seven-seat light utility helicopter.
Power Plant: One 500 shp Allison 250-C28B turboshaft.
Performance: (At 3,900 lb/1 769 kg) Max. speed, 144 mph (232 km/h); cruise, 136 mph (229 km/h) at sea level; hovering ceiling (in ground effect), 8,200 ft (2 499 m), (out of ground effect), 2,000 ft (610 m); range, 390 mls (628 km) at sea level, 430 mls (692 km) at 5,000 ft (1 524 m).
Weights: Empty, 2,160 lb (980 kg); max. take-off, 4,150 lb (1 882 kg).
Dimensions: Rotor diam, 37 ft 0 in (11,28 m); fuselage length, 33 ft 3 in (10,13 m).
Notes: The Model 206L-1 LongRanger II is a stretched and more powerful version of the Model 206B JetRanger III, with longer fuselage, increased fuel capacity, an uprated engine and a larger rotor. The LongRanger is being manufactured in parallel with the JetRanger III and initial customer deliveries commenced in October 1975, prototype testing having been initiated on September 11, 1974. The LongRanger is available with emergency flotation gear and with a 2,000-lb (907-kg) capacity cargo hook. In the aeromedical or rescue role the LongRanger can accommodate two casualty stretchers and two ambulatory casualties. The 206L-1 LongRanger II was introduced in 1978, and production was 15 monthly at the beginning of 1981, with more than 600 LongRangers delivered.

BELL MODEL 206L TEXASRANGER

Country of Origin: USA.
Type: Multi-role military helicopter.
Power Plant: One 650 shp Allison 250-C30P turboshaft.
Performance: (Utility configuration) Max. cruising speed, 131 mph (211 km/h); econ. cruise, 129 mph (207 km/h); max. inclined climb, 1,360 ft/min (6,9 m/sec); hovering ceiling (in ground effect), 12,000 ft (3 658 m); max. range (standard fuel), 356 mls (573 km).
Weights: Max. take-off (internal load), 4,150 lb (1 882 kg), (external jettisonable load), 4,250 lb (1 928 kg).
Dimensions: Rotor diam, 37 ft 0 in (11,28 m); fuselage length, 33 ft 3 in (10,13 m).
Notes: The TexasRanger is a multi-role military version of the Model 206L-1 LongRanger II (see page 225) introduced in 1980. The TexasRanger can carry four TOW (Tube-launched Optically-tracked Wire-guided) missiles for the anti-armour role, or two twin 7,62-mm gun pods with 500 rpg, or two pods each containing seven 2·75-in (70-mm) rockets. Armoured crew seats are fitted and missile control electronics are housed in the rear cabin. With the weapons pallet removed, the Texas-Ranger can carry seven persons, including the crew, and the helicopter is configured for quick-change to cover a variety of roles, including troop lift, command control, armed reconnais-sance and surveillance.

BELL AH-1S HUEYCOBRA

Country of Origin: USA.
Type: Two-seat light attack helicopter.
Power Plant: One 1,800 shp Avco Lycoming T53-L-703 turboshaft.
Performance: Max. speed, 172 mph (277 km/h), (TOW configuration), 141 mph (227 km/h); max. inclined climb, 1,620 ft/min (8,23 m/sec); hovering ceiling TOW configuration (in ground effect), 12,200 ft (3 720 m); max. range, 357 mls (574 km).
Weights: (TOW configuration) Operational empty, 6,479 lb (2 939 kg); max. take-off, 10,000 lb (4 535 kg).
Dimensions: Rotor diam, 44 ft 0 in (13,41 m); fuselage length, 44 ft 7 in (13,59 m).
Notes: The AH-1S is a dedicated attack and anti-armour helicopter serving primarily with the US Army which will have received 297 new-production AH-1S HueyCobras by mid-1981, plus 290 resulting from the conversion of earlier AH-1G and AH-1Q HueyCobras. Current planning calls for conversion of a further 372 AH-1Gs to AH-1S standards, and both conversion and new-production AH-1S HueyCobras are being progressively upgraded to "Modernised AH-1S" standard, the entire programme being scheduled for completion in 1985, resulting in a total of 959 "Modernised" AH-1S HueyCobras. In December 1979, one YAH-1S was flown with a four-bladed main rotor as the Model 249.

BELL AH-1T SEACOBRA

Country of Origin: USA.

Type: Two-seat light attack helicopter.

Power Plant: One 1,970 shp Pratt & Whitney T400-WV-402 coupled turboshaft.

Performance: (Attack configuration at 12,401 lb/5 625 kg) Max. speed, 181 mph (291 km/h) at sea level; average cruise, 168 mph (270 km/h); max. inclined climb, 2,190 ft/min (11,12 m/sec); hovering ceiling (out of ground effect), 5,350 ft (1 630 m); range, 276 mls (445 km).

Weights: Empty, 8,030 lb (3 642 kg); max. take-off, 14,000 lb (6 350 kg).

Dimensions: Rotor diam, 48 ft 0 in (14,63 m); fuselage length, 45 ft 3 in (13,79 m).

Notes: The SeaCobra is a twin-turboshaft version of the HueyCobra (see page 227), the initial model for the US Marine Corps having been the AH-1J (69 delivered of which two modified as AH-1T prototypes). The AH-1T features uprated components for significantly increased payload and performance, the first example having been delivered to the US Marine Corps on October 15, 1977, and a further 56 being delivered to the service of which 23 being modified to TOW configuration. The AH-1T has a three-barrel 20 mm cannon barbette under the nose, and four stores stations under the stub wings for seven- or 19-tube launchers, Minigun pods, etc.

BELL MODEL 214ST

Country of Origin: USA.
Type: Medium transport helicopter (19 seats).
Power Plant: Two 1,625 shp (limited to combined output of 2,250 shp) General Electric CT7-2 turboshafts.
Performance: Max. cruising speed, 164 mph (264 km/h) at sea level, 161 mph (259 km/h) at 4,000 ft (1 220 m); hovering ceiling (in ground effect), 12,600 ft (3 840 m), (out of ground effect), 3,300 ft (1 005 m); range (standard fuel), 460 mls (740 km).
Weights: Max. take-off (internal load), 15,500 lb (7 030 kg), (external jettisonable load), 16,500 lb (7 484 kg).
Dimensions: Rotor diam, 52 ft 0 in (15,85 m); fuselage length, 50 ft 0 in (15,24 m).
Notes: The Model 214ST (Super Transport) is a significantly improved derivative of the Model 214B BigLifter (see 1978 edition), production of which was phased out early 1981, and initial customer deliveries are scheduled for early 1982. The Model 214ST test-bed was first flown in March 1977, and the first of three representative prototypes (one in military configuration and two for commercial certification) commenced its test programme in August 1979. Work on an initial series of 100 helicopters of this type was to commence in 1981, with production tempo scheduled to reach three monthly in 1982. Alternative layouts are available for either 16 or 17 passengers.

BELL MODEL 222

Country of Origin: USA.
Type: Eight/ten-seat light utility and transport helicopter.
Power Plant: Two 620 shp Avco Lycoming LTS 101-650C-2 turboshafts.
Performance: Max. cruising speed, 150 mph (241 km/h) at sea level, 146 mph (235 km/h) at 8,000 ft (2 400 m); max. climb, 1,730 ft/min (8,8 m/sec); hovering ceiling (in ground effect), 10,300 ft (3 135 m), (out of ground effect), 6,400 ft (1 940 m); range (no reserves), 450 mls (724 km) at 8,000 ft (2 400 m).
Weights: Empty equipped, 4,577 lb (2 076 kg); max. take-off (standard configuration), 7,650 lb (3 470 kg).
Dimensions: Rotor diam, 39 ft 9 in (12,12 m); fuselage length, 39 ft 9 in (12,12 m).
Notes: The first of five prototypes of the Model 222 was flown on August 13, 1976, an initial production series of 250 helicopters of this type being initiated in 1978, with production deliveries commencing in January 1980, and some 55 delivered by beginning of 1981, when orders totalled approximately 180. Several versions of the Model 222 are on offer or under development, these including an executive version with a flight crew of two and five or six passengers and the so-called "offshore" model with accommodation for eight passengers and a flight crew of two. Options include interchangeable skids.

BELL MODEL 412

Country of Origin: USA.

Type: Fifteen-seat utility transport helicopter.

Power Plant: One 1,800 shp (1,308 shp take-off rating) Pratt & Whitney PT6T-3B turboshaft.

Performance: Max. speed, 149 mph (240 km/h) at sea level; cruise, 143 mph (230 km/h) at sea level, 146 mph (235 km/h) at 5,000 ft (1 525 m); hovering ceiling (in ground effect), 10,800 ft (3 290 m), (out of ground effect), 7,100 ft (2 165 m) at 10,500 lb/4 763 kg; max. range, 282 mls (454 km), (with auxiliary tanks), 518 mls (834 km).

Weights: Empty equipped, 6,070 lb (2 753 kg); max. take-off, 11,500 lb (5 216 kg).

Dimensions: Rotor diam, 46 ft 0 in (14,02 m); fuselage length, 41 ft 8½ in (12,70 m).

Notes: The Model 412, flown for the first time in August 1979, is an updated Model 212 (production of which was completed early 1981) with a new-design four-bladed rotor, a shorter rotor mast assembly, and uprated engine and transmission systems, giving more than twice the life of the Model 212 units. Composite rotor blades are used and the rotor head incorporates elastomeric bearings and dampers to simplify moving parts. A third prototype joined the test programme in 1980, an initial series of 200 helicopters being laid down with initial deliveries to commence in 1981, anticipated production rate building up to 100 annually by 1983.

BOEING VERTOL 234 CHINOOK

Country of Origin: USA.
Type: Commercial transport helicopter.
Power Plant: Two 4,075 shp Avco Lycoming AL 5512 turboshafts.
Performance: Max. cruising speed (at 47,000 lb/21 318 kg), 167 mph (269 km/h) at 2,000 ft (610 m); range cruise, 155 mph (250 km/h); max. inclined climb, 1,350 ft/min (6,8 m/sec); hovering ceiling (in ground effect), 9,150 ft (2 790 m), (out of ground effect), 4,900 ft (1 495 m); range (44 passengers and 45 min reserves), 627 mls (1 010 km), (max. fuel), 852 mls (1 371 km).
Weights: Empty, 24,449 lb (11 090 kg); max. take-off, 47,000 lb (21 318 kg).
Dimensions: Rotor diam (each), 60 ft 0 in (18,29 m); fuselage length, 52 ft 1 in (15,87 m).
Notes: Possessing an airframe based on the latest Model 414 military Chinook (see opposite), the Model 234 has been developed specifically for commercial purposes and two basic versions are offered, a long-range model described above and a utility model with fuel tank-housing side fairings removed. The first Model 234 was flown on August 19, 1980, certification is scheduled for mid-1981 and the first deliveries (to British Airways Helicopters with six on order) will follow during the course of the year, primarily for North Sea oil rig support.

BOEING VERTOL 414 CHINOOK HC Mk 1

Country of Origin: USA.
Type: Medium transport helicopter.
Power Plant: Two 3,750 shp Avco Lycoming T55-L-11E turboshafts.
Performance: (At 45,400 lb/20 593 kg) Max. speed, 146 mph (235 km/h) at sea level; average cruise, 131 mph (211 km/h); max. inclined climb, 1,380 ft/min (7,0 m/sec); service ceiling, 8,400 ft (2 560 m); max. ferry range, 1,190 mls (1 915 km).
Weights: Empty, 22,591 lb (10 247 kg); max. take-off, 50,000 lb (22 680 kg).
Dimensions: Rotor diam (each), 60 ft 0 in (18,29 m); fuselage length, 51 ft 0 in (15,55 m).
Notes: The Model 414 as supplied to the RAF as the Chinook HC Mk 1 combines some features of the US Army's CH-47D (see 1980 edition) and features of the Canadian CH-147, but with provision for glassfibre/carbonfibre rotor blades. The first of 33 Chinook HC Mk 1s for the RAF was flown on March 23, 1980 and accepted on December 2, 1980, with deliveries continuing through 1981. The RAF version can accommodate 44 troops and has three external cargo hooks. Boeing Vertol will manufacture 32 Chinooks (mostly HC Mk 1s for the RAF) during 1981, and will initiate the conversion to essentially similar CH-47D standards a total of 436 CH-47As, Bs and Cs.

HUGHES 500MD DEFENDER II

Country of Origin: USA.
Type: Light gunship and multi-role helicopter.
Power Plant: One 420 shp Allison 250-C20B turboshaft.
Performance: (At 3,000 lb/1 362 kg) Max. speed, 175 mph (282 km/h) at sea level; cruise, 160 mph (257 km/h) at 4,000 ft (1 220 m); max. inclined climb, 1,920 ft/min (9,75 m/sec); hovering ceiling (in ground effect), 8,800 ft (2 682 m), (out of ground effect), 7,100 ft (2 164 m); max. range, 263 mls (423 km).
Weights: Empty, 1,295 lb (588 kg): max. take-off (internal load), 3,000 lb (1 362 kg), (with external load), 3,620 lb (1 642 kg).
Dimensions: Rotor diam, 26 ft 5 in (8,05 m); fuselage length, 21 ft 5 in (6,52 m).
Notes: The Defender II multi-mission version of the Model 500MD was introduced mid-1980 for 1982 delivery, and features a Martin Marietta rotor mast-top sight, a General Dynamics twin-Stinger air-to-air missile pod, an underfuselage 30-mm chain gun and a pilot's night vision sensor. The Defender II can be rapidly reconfigured for anti-armour target designation, anti-helicopter, suppressive fire and transport roles. The Model 500MD TOW Defender (carrying four tube-launched optically-tracked wire-guided anti-armour missiles) is currently in service with Israel (30), South Korea (25) and Kenya (15).

HUGHES AH-64

Country of Origin: USA.
Type: Tandem two-seat attack helicopter.
Power Plant: Two 1,536 shp General Electric T700-GE-700 turboshafts.
Performance: Max. speed, 191 mph (307 km/h); cruise, 179 mph (288 km/h); max. inclined climb, 3,200 ft/min (16,27 m/sec); hovering ceiling (in ground effect), 14,600 ft (4 453 m), (outside ground effect), 11,800 ft (3 600 m); service ceiling, 8,000 ft (2 400 m); max. range, 424 mls (682 km).
Weights: Empty, 9,900 lb (4 490 kg); primary mission, 13,600 lb (6 169 kg); max. take-off, 17,400 lb (7 892 kg).
Dimensions: Rotor diam, 48 ft 0 in (14,63 m); fuselage length, 49 ft 4½ in (15,05 m).
Notes: Winning contender in the US Army's AAH (Advanced Attack Helicopter) contest, the YAH-64 flew for the first time on September 30, 1975. Two prototypes were used for the initial trials, the first of three more with fully integrated weapons systems commenced trials on October 31, 1979, a further three following in 1980. Planned total procurement comprises 536 AH-64s. The AH-64 is armed with a single-barrel 30-mm gun based on the chain-driven bolt system and suspended beneath the forward fuselage, and eight BGM-71A TOW anti-armour missiles may be carried, alternative armament including 16 Hellfire laser-seeking missiles. Target acquisition and designation and a pilot's night vision system will be used.

KAMOV KA-25 (HORMONE A)

Country of Origin: USSR.

Type: Shipboard anti-submarine warfare helicopter.

Power Plant: Two 900 shp Glushenkov GTD-3 turboshafts.

Performance: (Estimated) Max. speed, 130 mph (209 km/h); normal cruise, 120 mph (193 km/h); max. range, 400 mls (644 km); service ceiling, 11,000 ft (3 353 m).

Weights: (Estimated) Empty, 10,500 lb (4 765 kg); max. take-off, 16,500 lb (7 484 kg).

Dimensions: Rotor diam (each), 51 ft 7½ in (15,74 m); approx. fuselage length, 35 ft 6 in (10,82 m).

Notes: Possessing a basically similar airframe to that of the Ka-25K (see 1973 edition) and employing a similar self-contained assembly comprising rotors, transmission, engines and auxiliaries, the Ka-25 serves with the Soviet Navy primarily in the ASW role but is also employed in the utility and transport roles. The ASW Ka-25 serves aboard the helicopter cruisers *Moskva* and *Leningrad*, and the carriers *Kiev* and *Minsk*, as well as with shore-based units. A search radar installation is mounted in a nose randome, but other sensor housings and antennae differ widely from helicopter to helicopter. There is no evidence that externally-mounted weapons may be carried. Each landing wheel is surrounded by an inflatable pontoon surmounted by inflation bottles. The Hormone-A is intended for ASW operations whereas the Hormone-B is used for over-the-horizon missile targeting.

MBB BO 105L

Country of Origin: Federal Germany.
Type: Five/six-seat light utility helicopter.
Power Plant: Two 550 shp Allison 250-C28C turboshafts.
Performance: Max. speed, 168 mph (270 km/h) at sea level; max. cruise, 157 mph (252 km/h) at sea level; max. climb, 1,970 ft/min (10 m/sec); hovering ceiling (in ground effect), 13,120 ft (4 000 m), (out of ground effect), 11,280 ft (3 440 m); range, 286 mls (460 km).
Weights: Empty, 2,756 lb (1 250 kg); max. take-off, 5,291 lb (2 400 kg), (with external load), 5,512 lb (2 500 kg).
Dimensions: Rotor diam, 32 ft 3½ in (9,84 m); fuselage length, 28 ft 1 in (8,56 m).
Notes: The BO 105L is a derivative of the BO 105CB (see 1979 edition) with uprated transmission and more powerful turboshaft for "hot-and-high" conditions. It is otherwise similar to the BO 105CB (420 shp Allison 250-C20B) which was continuing in production at the beginning of 1981, when more than 600 BO 105s (all versions) had been delivered and production was running at 10–12 monthly, and licence assembly was being undertaken in Indonesia, the Philippines and Spain. Deliveries to the Federal German Army of 227 BO 105M helicopters for liaison and observation tasks commenced late 1979, and deliveries of 212 HOT-equipped BO 105s for the antiarmour role began on December 4, 1980. The latter have uprated engines and transmission systems.

MBB-KAWASAKI BK 117

Countries of Origin: Federal Germany and Japan.
Type: Multi-purpose eight-to-twelve-seat helicopter.
Power Plant: Two 600 shp Avco Lycoming LTS 101-650B-1 turboshafts.
Performance: Max. speed, 171 mph (275 km/h) at sea level; cruise, 164 mph (264 km/h) at sea level; max. climb, 1,970 ft/min (10 m/sec); hovering ceiling (in ground effect), 13,450 ft (4 100 m), (out of ground effect), 10,340 ft (3 150 m); range (max. payload), 339 mls (545,4 km).
Weights: Empty, 3,351 lb (1 520 kg); max. take-off, 6,173 lb (2 800 kg).
Dimensions: Rotor diam, 36 ft 1 in (11,00 m); fuselage length, 32 ft 5 in (9,88 m).
Notes: The BK 117 is a co-operative development between Messerschmitt-Bolkow-Blohm and Kawasaki, the first of two flying prototypes commencing its flight test programme on June 13, 1979 (in Germany), with the second following on August 10 (in Japan). A decision to proceed with series production was taken in 1980, with production deliveries commencing first quarter of 1982. Some 120 BK 117s had been ordered by the beginning of 1981. MBB is responsible for the main and tail rotor systems, tail unit and hydraulic components, while Kawasaki is responsible for production of the fuselage, undercarriage, transmission and some other components. Several military versions are currently proposed.

MIL MI-8 (HIP)

Country of Origin: USSR.

Type: Assault transport helicopter.

Power Plant: Two 1,700 shp Isotov TV2-117A turboshafts.

Performance: Max. speed, 161 mph (260 km/h) at 3,280 ft (1 000 m), 155 mph (250 km/h) at sea level; max. cruise, 140 mph (225 km/h); hovering ceiling (in ground effect), 6,233 ft (1 900 m), (out of ground effect), 2,625 ft (800 m); range (standard fuel), 290 mls (465 km).

Weights: (Hip-C) Empty, 14,603 lb (6 624 kg); normal loaded, 24,470 lb (11 100 kg); max. take-off, 26,455 lb (12 000 kg).

Dimensions: Rotor diam, 69 ft 10$\frac{1}{4}$ in (21,29 m); fuselage length, 60 ft 0$\frac{3}{4}$ in (18,31 m).

Notes: Currently being manufactured at a rate of 700–800 annually, with 6,000–7,000 delivered for civil and military use since its debut in 1961, the Mi-8 is numerically the most important Soviet helicopter. Current military versions include the Hip-C basic assault transport, the Hip-D with additional antennae and podded equipment for electronic tasks, the Hip-E and the Hip-F, the former carrying up to six rocket pods and four Swatter IR-homing anti-armour missiles, and the latter carrying six Sagger wire-guided anti-armour missiles. The Mi-8 can accommodate 24 troops or 12 stretchers, and most have a 12,7-mm machine gun in the nose. Commercial models include the basic 28–32 passenger model and the Mi-8T utility version.

MIL MI-14 (HAZE-A)

Country of Origin: USSR.
Type: Amphibious anti-submarine helicopter.
Power Plant: Two 1,500 shp Isotov TV-2 turboshafts.
Performance: (Estimated) Max. speed, 143 mph (230 km/h); max. cruise, 130 mph (210 km/h); hovering ceiling (in ground effect), 5,250 ft (1 600 m), (out of ground effect), 2,295 ft (700 m); tactical radius, 124 mls (200 km).
Weights: (Estimated) Max. take-off, 26,455 lb (12 000 kg).
Dimensions: Rotor diam, 69 ft 10¼ in (21,29 m); fuselage length, 59 ft 7 in (18,15 m).
Notes: The Mi-14 amphibious anti-submarine warfare helicopter, which serves with shore-based elements of the Soviet Naval Air Force, is a derivative of the Mi-8 (see page 239) with essentially similar power plant and dynamic components, and much of the structure is common between the two helicopters. New features include the boat-type hull, outriggers which, housing the retractable lateral twin-wheel undercarriage members, incorporate water rudders, a search radar installation beneath the nose and a sonar "bird" beneath the tailboom root. The Mi-14 may presumably be used for over-the-horizon missile targeting and for such tasks as search and rescue. It may also be assumed that the Mi-14 possesses a weapons bay for ASW torpedoes, nuclear depth charges and other stores. This amphibious helicopter reportedly entered service in 1975.

MIL MI-24 (HIND-D)

Country of Origin: USSR.

Type: Assault and anti-armour helicopter.

Power Plant: Two 2,200 shp Isotov TV3-117 turboshafts.

Performance: (Estimated) Max. speed, 170–180 mph (273–290 km/h) at 3,280 ft (1 000 m); max. cruise, 145 mph (233 km/h); max. inclined climb rate, 3,000 ft/min (15,24 m/sec).

Weights: (Estimated) Normal take-off, 22,000 lb (10 000 kg).

Dimensions: (Estimated) Rotor diam, 55 ft 0 in (16,76 m); fuselage length, 55 ft 6 in (16,90 m).

Notes: By comparison with the Hind-A version of the Mi-24 (see 1977 edition), the Hind-D embodies a redesigned forward fuselage and is optimised for the gunship role, having tandem stations for the weapons operator (in nose) and pilot. The Hind-D can accommodate eight fully-equipped troops, has a barbette-mounted four-barrel rotary-type 12,7-mm cannon beneath the nose and can carry up to 2,800 lb (1 275 kg) of ordnance externally, including four AT-2 Swatter IR-homing anti-armour missiles and four pods each with 32 57-mm rockets. It has been exported to Afghanistan, Algeria, Bulgaria, Czechoslovakia, East Germany, Hungary, Iraq, Libya, Poland and South Yemen. The Hind-E is similar but has provision for four laser-homing tube-launched Spiral anti-armour missiles, and embodies some structural hardening, steel and titanium being substituted for aluminium in certain critical components. More than 500 Hind-D and -E deployed by WarPac.

PZL KANIA

Country of Origin: Poland.

Type: Eight/ten-seat light utility and transport helicopter.

Power Plant: Two 420 shp Allison 250-C20B turboshafts.

Performance: Max. cruising speed, 130 mph (210 km/h) at sea level; econ. cruise, 121 mph (195 km/h); max. inclined climb, 1,476 ft/min (7,5 m/sec); hovering ceiling (in ground effect), 6,365 ft (1 940 m), (out of ground effect), 4,070 ft (1 240 m); range (standard fuel), 317 mls (510 km).

Weights: Empty equipped, 4,717 lb (2 140 kg); max. take-off, 7,826 lb (3 550 kg).

Dimensions: Rotor diam, 47 ft $9\frac{1}{4}$ in (14,56 m); fuselage length, 39 ft $2\frac{1}{2}$ in (11,95 m).

Notes: The Kania (also named Kitty Hawk for the western export market) is a derivative of the licence-built Mil Mi-2, the first of two prototypes flying on June 3, 1979, and production deliveries being scheduled for the first half of 1981. A second version of the Kania intended specifically for the US market is to be known as the Taurus, this having more powerful Allison 250-C28 turboshafts, with a redesigned single-orifice air intake cowling and revised nose contours. The Kania is intended for agricultural, aeromedical and other roles, and will succeed the Mi-2.

242

PZL W-3 SOKOL

Country of Origin: Poland.
Type: Twelve-passenger transport helicopter.
Power Plant: Two 1,000 shp PZL-10W (Glushenkov TVD-10) turboshafts.
Performance: (Estimated) Max. speed, 158 mph (255 km/h); cruise, 137 mph (220 km/h); max. inclined climb, 1,850 ft/min (9,4 m/sec); hovering ceiling (out of ground effect), 5,825 ft (1 775 m); service ceiling, 16,405 ft (5 000 m); range (standard fuel), 357 mls (575 km).
Weights: Operational empty, 5,489 lb (2 490 kg); max. take-off, 13,227 lb (6 000 kg).
Dimensions: Rotor diam, 51 ft 6 in (15,70 m); fuselage length, 44 ft 5 in (13,54 m).
Notes: The Sokol (Falcon) was flown for the first time on November 16, 1979, and is intended to fill a size gap between the licence-built Mil Mi-2 and the Mil Mi-8. Certification of the prototype is expected during the course of 1981, and production deliveries are envisaged from late 1982 or early 1983, passenger, freight, aeromedical and agricultural versions being planned. Two crew members are accommodated side-by-side on the flight deck, and in the ambulance role four stretchers will be accommodated.

SIKORSKY S-61D (SEA KING)

Country of Origin: USA.

Type: Amphibious anti-submarine helicopter.

Power Plant: Two 1,500 shp General Electric T58-GE-10 turboshafts.

Performance: Max. speed, 172 mph (277 km/h) at sea level; inclined climb, 2,200 ft/min (11,2 m/sec); hovering ceiling (out of ground effect), 8,200 ft (2 500 m); range (with 10% reserves), 622 mls (1 000 km).

Weights: Empty equipped, 12,087 lb (5 481 kg); max. take-off, 20,500 lb (9 297 kg).

Dimensions: Rotor diam, 62 ft 0 in (18,90 m); fuselage length, 54 ft 9 in (16,69 m).

Notes: A more powerful derivative of the S-61B, the S-61D serves with the US Navy, as the SH-3D, 72 helicopters of this type following on production of 255 SH-3As (S-61Bs) for the ASW role for the US Navy, four being supplied to the Brazilian Navy and 22 to the Spanish Navy. Four similar aircraft have been supplied to the Argentine Navy as S-61D-4s and 11 have been supplied to the US Army/US Marine Corps Executive Flight Detachment as VH-3Ds. Licence manufacture of the S-61D is being undertaken in the United Kingdom (see page 250), in Japan for the Maritime Self-Defence Force and in Italy by Agusta for the Italian, Iranian and Peruvian (illustrated) navies. The SH-3G and SH-3H are upgraded versions of the SH-3A.

SIKORSKY S-61R

Country of Origin: USA

Type: Amphibious transport and rescue helicopter.

Power Plant: (CH-3E) Two 1,500 shp General Electric T58-GE-5 turboshafts.

Performance: (CH-3E at 21,247 lb/9 635 kg) Max. speed, 162 mph (261 km/h) at sea level; range cruise, 144 mph (232 km/h); max. inclined climb, 1,310 ft/min (6,6 m/sec); hovering ceiling (in ground effect), 4,100 ft (1 250 m); range (with 10% reserves), 465 mls (748 km).

Weights: (CH-3E) Empty, 13,255 lb (6 010 kg); normal take-off, 21,247 lb (9 635 kg); max. take-off, 22,050 lb (10 000 kg).

Dimensions: Rotor diam, 62 ft 0 in (18,90 m); fuselage length, 57 ft 3 in (17,45 m).

Notes: Although based on the S-61A, the S-61R embodies numerous design changes, including a rear ramp and a tricycle-type undercarriage. Initial model for the USAF was the CH-3C with 1,300 shp T58-GE-1 turboshafts, but this was subsequently updated to CH-3E standards. The CH-3E can accommodate 25–30 troops or 5,000 lb (2 270 kg) of cargo, and may be fitted with a TAT-102 barbette on each sponson mounting a 7,62-mm Minigun. The HH-3E is a USAF rescue version with armour, self-sealing tanks, and refuelling probe, and the HH-3F Pelican is a US Coast Guard search-and-rescue model. The HH-3F version of the S-61R is built under licence in Italy by Agusta.

SIKORSKY CH-53E SUPER STALLION

Country of Origin: USA.

Type: Amphibious assault transport helicopter.

Power Plant: Three 4,380 shp General Electric T64-GE-415 turboshafts.

Performance: (At 56,000 lb/25 400 kg) Max. speed, 196 mph (315 km/h) at sea level; cruise, 173 mph (278 km/h) at sea level; max. inclined climb, 2,750 ft/min (13,97 m/sec); hovering ceiling (in ground effect), 11,550 ft (3 520 m), (out of ground effect), 9,500 ft (2 895 m); range, 1,290 mls (2 075 km).

Weights: Empty, 32,878 lb (14 913 kg); max. take-off, 73,500 lb (33 339 kg).

Dimensions: Rotor diam, 79 ft 0 in (24,08 m); fuselage length, 73 ft 5 in (22,38 m).

Notes: The CH-53E is a growth version of the CH-53D Sea Stallion (see 1974 edition) embodying a third engine, an uprated transmission system, a seventh main rotor blade and increased rotor diameter. The first of two prototypes was flown on March 1, 1974, and the first of two pre-production examples followed on December 8, 1975, successive production orders totalling 49 helicopters to be divided between the US Navy (16) and US Marine Corps (33), the first production Super Stallion having made its first flight on December 13, 1980. The CH-53E can accommodate up to 55 troops in a high-density seating arrangement. Fleet deliveries will begin mid-1981.

SIKORSKY S-70 (UH-60A) BLACK HAWK

Country of Origin: USA.

Type: Tactical transport helicopter.

Power Plant: Two, 1,543 shp General Electric T700-GE-700 turboshafts.

Performance: Max. speed, 224 mph (360 km/h) at sea level; cruise, 166 mph (267 km/h); vertical climb rate, 450 ft/min (2,28 m/sec); hovering ceiling (in ground effect), 10,000 ft (3 048 m), (out of ground effect), 5,800 ft (1 758 m); endurance, 2·3–3·0 hrs.

Weights: Design gross, 16,500 lb (7 485 kg); max. take-off, 22,000 lb (9 979 kg).

Dimensions: Rotor diam, 53 ft 8 in (16,23 m); fuselage length, 50 ft 0¾ in (15,26 m).

Notes: The Black Hawk was winner of the US Army's UTTAS (Utility Tactical Transport Aircraft System) contest, and contracts had been announced by beginning of 1981 for 255 examples. The first of three YUH-60As was flown on October 17, 1974, and a company-funded fourth prototype flew on May 23, 1975. The Black Hawk is primarily a combat assault squad carrier, accommodating 11 fully-equipped troops, but it is capable of carrying an 8,000-lb (3 629-kg) slung load. Two variants under development at the beginning of 1981 were the EC-60A ECM model and EC-60B for target acquisition. The first production deliveries to the US Army were made in June 1979, with 88 delivered by mid-November 1980.

SIKORSKY S-70L (SH-60B) SEAHAWK

Country of Origin: USA.

Type: Shipboard multi-role helicopter.

Power Plant: Two 1,690 shp General Electric T700-GE-401 turboshafts.

Performance: (At 20,244 lb/9 183 kg) Max. speed, 167 mph (269 km/h) at sea level; max. cruising speed, 155 mph (249 km/h) at 5,000 ft (1 525 m); max. vertical climb, 1,192 ft/min (6,05 m/sec); time on station (at radius of 57 mls/92 km), 3 hrs 52 min.

Weights: Empty equipped, 14,019 lb (6 359 kg); max. take-off, 21,844 lb (9 908 kg).

Dimensions: Rotor diam, 53 ft 8 in (16,36 m); fuselage length, 50 ft 0¾ in (15,26 m).

Notes: Winner of the US Navy's LAMPS (Light Airborne Multi-Purpose System) Mk III helicopter contest, the SH-60B is intended to fulfil both anti-submarine warfare (ASW) and anti-ship surveillance and targeting (ASST) missions, and the first of five prototypes was flown on December 12, 1979, and the last on July 14, 1980. Evolved from the UH-60A (see page 247), the SH-60B is intended to serve aboard DD-963 destroyers, DDG-47 Aegis cruisers and FFG-7 guided-missile frigates as an integral extension of the sensor and weapon system of the launching vessel. Shipboard trials were scheduled to be undertaken in January 1981, and the US Navy has a requirement for 204 LAMPS III category helicopters with deliveries from 1984.

SIKORSKY S-76

Country of Origin: USA.

Type: Fourteen-seat commercial transport helicopter.

Power Plant: Two 700 shp Allison 250-C30 turboshafts.

Performance: Max. speed, 179 mph (288 km/h); max. cruise, 167 mph (268 km/h); range cruise, 145 mph (233 km/h); hovering ceiling (in ground effect), 5,100 ft (1 524 m), (out of ground effect), 1,400 ft (427 m); range (full payload and 30 min reserve), 460 mls (740 km).

Weights: Empty, 4,942 lb (2 241 kg); max. take-off, 9,700 lb (4 399 kg).

Dimensions: Rotor diam, 44 ft 0 in (13,41 m); fuselage length, 44 ft 1 in (13,44 m).

Notes: The first of four prototypes of the S-76 flew on March 13, 1977, and customer deliveries commenced 1979, with 40 being delivered by the beginning of 1980, when a production rate of eight per month had been attained. The S-76 is unique among Sikorsky commercial helicopters in that conceptually it owes nothing to an existing military model, although it has been designed to conform with appropriate military specifications and military customers were included among contracts for 426 helicopters of this type that had been ordered by the beginning of 1981, when a total of 125 had been delivered. The S-76 may be fitted with extended-range tanks, cargo hook and rescue hoist. The main rotor is a scaled-down version of that used by the UH-60.

WESTLAND SEA KING

Country of Origin: United Kingdom (US licence).

Type: Anti-submarine warfare and search-and-rescue helicopter.

Power Plant: Two, 1,060 shp Rolls-Royce Gem 41-1 turboshafts.

Performance: Max. speed, 143 mph (230 km/h); max. continuous cruise at sea level, 131 mph (211 km/h); hovering ceiling (in ground effect), 5,000 ft (1 525 m) (out of ground effect), 3,200 ft (975 m); range (standard fuel), 764 mls (1 230 km), (auxiliary fuel), 937 mls (1 507 km).

Weights: Empty equipped (ASW), 13,672 lb (6 201 kg), (SAR), 12,376 lb (5 613 kg); max. take-off, 21,000 lb (9 525 kg).

Dimensions: Rotor diam, 62 ft 0 in (18,90 m); fuselage length, 55 ft 9¾ in (17,01 m).

Notes: The Sea King Mk 2 is an uprated version of the basic ASW and SAR derivative of the licence-built S-61D (see page 244), the first Mk 2 being flown on June 30, 1974, and being one of 10 Sea King Mk 50s ordered by the Australian Navy. Twenty-one delivered to the Royal Navy as Sea King HAS Mk 2s, and 15 examples of a SAR version to the RAF as Sea King HAR Mk 3s. Current production version is the Sea King HAS Mk 5 (illustrated), delivery of 17 to Royal Navy having commenced October 1980. All HAS Mk 2s will be brought up to Mk 5 standards. A total of 239 Westland-built derivatives of the S-61D had been ordered by the beginning of 1981.

WESTLAND WG 13 LYNX

Country of Origin: United Kingdom.
Type: Multi-purpose, ASW and transport helicopter.
Power Plant: Two 900 shp Rolls-Royce BS.360-07-26 Gem 100 turboshafts.
Performance: Max. speed, 207 mph (333 km/h); max. continuous sea level cruise, 170 mph (273 km/h); max. inclined climb, 1,174 ft/min (11,05 m/sec); hovering ceiling (out of ground effect), 12,000 ft (3 660 m); max. range (internal fuel), 391 mls (629 km); max. ferry range (auxiliary fuel), 787 mls (1 266 km).
Weights: (HAS Mk 2) Operational empty, 6,767–6,999 lb (3 069–3 174 kg); max. take-off, 9,500 lb (4 309 kg).
Dimensions: Rotor diam, 42 ft 0 in (12,80 m); fuselage length, 39 ft 1¼ in (11,92 m).
Notes: The first of 13 development Lynxes was flown on March 21, 1971, with the first production example (an HAS Mk 2) flying on February 10, 1976. By the beginning of 1981 production rate was nine per month and 306 were on order, including 40 for the French Navy, 80 for the Royal Navy, 114 for the British Army, 10 for the Argentine Navy, eight for the Danish Navy, 12 for the German Navy, nine for the Brazilian Navy, six for Norway, 24 for the Netherlands Navy and three of a general-purpose version for Qatar. The Lynx AH Mk 1 is the British Army's general utility version and the Lynx HAS Mk 2 is the ASW version for the Royal Navy. Eighteen of the Dutch and 14 of the French Lynx have uprated engines.

WESTLAND WG 30

Country of Origin: United Kingdom.
Type: Transport and utility helicopter.
Power Plant: Two 1,060 shp Rolls-Royce Gem 41-1 turbo-shafts.
Performance: Max. speed (at 10,500 lb/4 763 kg), 163 mph (263 km/h) at 3,000 ft (915 m); hovering ceiling (in ground effect), 7,200 ft (2 195 m), (out of ground effect), 5,000 ft (1 525 m); range (seven passengers), 426 mls (686 km).
Weights: Operational empty (typical), 6,880 lb (3 120 kg); max. take-off, 11,750 lb (5 330 kg).
Dimensions: Rotor diam, 43 ft 8 in (13,31 m); fuselage length, 47 ft 0 in (14,33 m).
Notes: The WG 30, flown for the first time on April 10, 1979, is a private venture development of the Lynx (see page 251) featuring an entirely new fuselage offering a substantial increase in capacity. Aimed primarily at the multi-role military helicopter field, the WG 30 has a crew of two and in the transport role can carry 17–22 passengers. Commitment to the WG 30 at the time of closing for press covers the flight development of two prototypes aimed at certification during 1981, and an initial production batch of 20 delivery from mid-1982, British Airways have indicated intention to procure a quantity. The first production WG 30 is scheduled to be completed in October 1981. The WG 30 utilises more than 85% of the proven systems of the WG 13 Lynx.

ACKNOWLEDGEMENTS

The author wishes to record his thanks to the many aircraft manufacturers that have supplied information and photographs for inclusion in this volume, and to the following sources of copyright photographs: Air Portraits, pages 16 and 18; Austin J. Brown, pages 14 and 206; Jan Čech, page 128; Flug Revue International, page 156; P. A. Jackson, page 60; Andrew March, page 180, and the Swedish Air Force, page 204. The three-view silhouette drawings published in this volume are copyright Pilot Press Limited and may not be reproduced without prior permission

INDEX OF AIRCRAFT TYPES